Stephen Hawking's Universe

THE

COSMOS

EXPLAINED

The Big Bang,

black holes,

white dwarfs,

time warps,

life,

the universe

and everything;

all explained in everyday language

Stephen Hawking's Universe

THE COSMOS EXPLAINED

DAVID FILKIN

BasicBooks
A Division of HarperCollins*Publishers*

For the brightest stars in my own universe: Neil, Jonathan and Matthew

FIRST EDITION

Designed by Elliott Beard.

Library of Congress Cataloging-in-Publication Data

Filkin, David.
Stephen Hawking's universe : the cosmos explained / David Filkin. —1st ed.
        p.   cm.
Includes index.
ISBN 0-465-08199-1
1. Cosmology.   2. Hawking, S. W. (Stephen W.) Brief history of time.   I. Title.
QB981.F48   1997
523.1—dc21   97-33121
                                                            CIP

97 98 99 00 01 ❖/RRD 10 9 8 7 6 5 4 3 2 1

Contents

Contents

Acknowledgments

IT WOULD BE ARROGANT, IF NOT ABSURD, TO PRETEND THAT I, SINGLE-handed, could ever have attempted to put together a complete picture of the nature of the universe—let alone a compelling explanation easily understood by everyone. I began with far too little knowledge of such exotic ideas as white dwarfs, the Big Bang and black holes. I was even—what heresy—more than a little suspicious that they might be figments of eccentric academic fantasy, rather than essential pieces in a scientific jigsaw. It does not help most of us to have faith in this jigsaw picture when it turns out to look like nothing we have ever experienced. But I had an unquenchable desire to learn more, and a conviction that millions of others wanted to understand these things too. So I found myself persuading a fellow student from the early 1960s, Stephen Hawking, that I should make a television series—and subsequently write this book—based on what he had written in *A Brief History of Time*. The phenomenal success of Stephen's book has confirmed that there is a worldwide fascination with cosmology, even if many of his readers sheepishly confess that they have not quite understood everything. All this quickly convinced me that there was still room for another approach.

It is very much to the credit of Michael Jackson, then the Controller of BBC 2, that he trusted me to find my way through the seemingly impenetrable maze of modern physics, and to come up with six television documentaries as a result. With the patient help of Simon Singh, who really did understand the science, I prepared a proposal with which to woo coproducers and win funding. Brian Whitt, a colleague of Stephen Hawking's who had been closely involved in producing *A Brief History of Time*, spent long hours helping me refine the ideas before they went to press. Michael Attwell, then the Commissioning Editor for Factual Programmes at the BBC, and Bill Grant in New York were sufficiently captivated by them for the BBC and WNET in the U.S.A. to agree to coproduce the series. I am indebted to them all. Without their faith and encouragement, I would never have got started.

Patrick Uden, William Miller and Mary Phelps gave me office space at Uden Associates while I was producing the series, and we recruited an impressively talented production team. Between them, Philip Martin, Steve Davis, Joanna Haywood, Dan Gluckman, Kate Cox, Jessica Whitehead and Katie Gwyn managed to gather the facts and mix in the magic that turns solid data into informative entertainment. Without them I could not have made the television programs, and I probably would not have learned enough to begin writing this book. I cannot thank them enough.

I must not forget Sue Masey and all those working with Stephen Hawking, who have responded to every request for help promptly, imaginatively and generously. And above all, there is Stephen himself. He is alarmingly dismissive of his own importance. I sent him some notes for his corrections and thoughts, among which I had written ". . . Stephen Hawking, acknowledged as one of the leading authorities on black holes." The tightly typed pages were returned with Sue Masey's clear handwritten annotations in the margin, faithfully reproducing Stephen's thoughts. My line had been neatly scored out and an alternative added: ". . . Stephen Hawking, who has contributed to the study of black holes."

However much he may seek to play down his own significance, I am in no doubt that Stephen has been a remarkable inspiration for me. My gratitude to him cannot be expressed adequately.

Foreword

~~~~~~~~~

by Stephen Hawking

AS A YOUNG BOY, I WAS THE DESPAIR OF MY PARENTS BECAUSE I WAS always taking things apart to see what made them tick. Of course I generally couldn't get them back together again, but I felt that if I understood how something worked, then in a sense I was the master of it. I'm sure other people must feel the same: we find ourselves in a world that often doesn't seem to notice our existence or care about it and which sometimes is downright hostile. Yet if we can unscrew the front panel of the universe and look behind, we might be able to figure out how the little wheels work and feel that we have some control over what is going on. Fortunately, we aren't called upon to put the universe back together again.

Joking apart, I feel there's a fundamental point here. I feel helpless if I'm taken around a strange town of which I can't form a picture. We all need a mental map of the world showing where we are. Of course, our picture will have many layers of which the physical description is only one. Nevertheless, it is the bedrock on which all the other structures are based. If we have secured the foundations, we have a measure of control over the higher levels.

The physical laws that govern the universe are usually expressed in the form of mathematical equations. For most people, this has created a great barrier to understanding. But equations in physics are like the financial appendices to the budget: important if you are an accountant concerned with the details but unnecessary for a general understanding of what is going on. The basic ideas in physics can be explained in words and pictures. I personally don't like equations: it is hard to keep track of all the terms in one's head and I find it cumbersome to write them down (though I can do so on my computer using a language called TeX). I'm therefore always looking for ways to treat problems geometrically so I can see the answer in pictures, though that too has its difficulties: it is difficult enough to imagine objects in the three dimensions of space and one dimension of time that we are used to, let alone the seven or more extra hidden dimensions that may be there, according to our unified theories of everything. Still, one can generally ignore most of these dimensions and just picture things in the two or three dimensions that our brains are capable of visualizing. So I believe it is possible for everyone to understand the basic laws and forces that govern and shape the universe.

The laws of science that we have discovered show how one event is caused by another. It is natural to ask: What happens as one follows this causal chain of events back in time? Is there a first *cause* or does the chain go back forever? This is the chicken and egg question. The remarkable discovery that we have made this century is that there is indeed a first event, the Big Bang, which is maybe more like an egg than a chicken, though not really like either. At the Big Bang, the universe and time itself came into existence, so this is the first cause. If we could understand the Big Bang, we would know why the universe is the way it is. It used to be thought that it was impossible to apply the laws of science to the beginning of the universe, and indeed that it was sacrilegious to try. But recent developments in unifying the two pillars of twentieth-century science, Einstein's General Theory of Relativity and the Quantum Theory, have encouraged us to believe that it may be possible to find laws that hold even at the creation of the universe. In that case, everything in the universe would be determined by the laws of science. So if we understood those laws, we would in a sense be Masters of the Universe.

STEPHEN HAWKING
Cambridge, July 28, 1997

Stephen Hawking's Universe

THE

COSMOS

EXPLAINED

Introduction

~~~~~~

Finding a Boat
to Steer

It was not where we usually met. Eight of the University College, Oxford, rugby team, including myself, stood uneasily onboard the beautiful old college barge, waiting to try our hands at rowing for the first time. We were a strange mixture of long lanky line-out jumpers, embarrassingly fit and athletic three-quarters, decidedly sturdy and even more decidedly unfit front-row forwards, and a flanker. All we had in common were our blue and gold jerseys; yet we imagined, somehow, that someone was going to weld us into a race-winning crew.

It soon dawned on me that we were not alone. A much smaller figure stood alongside our group, distinguished by a blazer instead of a rugby

---

Syrius, the brightest star in the sky. The stars have always been used by sailors and travelers to guide them.

3

shirt, huge dark horn-rimmed spectacles, and an immaculate straw boater.

"Who's that?" I muttered to the person standing next to me.

"Hawking. Stephen Hawking," he whispered back. "He's going to be our cox."

"Bit of a playboy, actually," ventured someone else, "but fiendishly bright. Second-year physics."

I vaguely remembered having seen the boater across the main quadrangle and hearing his voice at dinner in the college hall. But I knew nothing much else about Stephen. Nor was I going to find out a lot more while we rowed. He was at one end of the boat; I was at the other. And there was no time for gossip. We had about three training outings before we had to race, and we had to learn everything in that time. I forget who our coach was; it does not really matter. He tried hard to give us some idea of what to do; but I think inwardly he, and all eight of us, knew we were not destined to perform with distinction. Only Stephen was prepared to think otherwise. He barked orders as he steered, refused to give up on any of us, and somehow convinced us by the day of our first race that we were not quite so hopeless after all.

On the narrow stretch of the river where racing takes place at Oxford, there is no room to race up a course side by side with each boat in its own lane. So the races are bumping races; each boat chases the boat ahead, and is in turn chased by the one behind. As soon as one boat catches up with another, the cox has to steer adroitly to make actual contact with the boat ahead, and so record a bump. Both crews then pull over to the side of the river, out of the race. On the next day of racing, the two boats swap places in the starting order; so, over four days of racing, a crew that does well can advance four places on the river.

We had inherited a position from the previous year's University College Rugby Eight that was well back down the river. At the starter's gun, Stephen had us set off at a suicidal pace. It certainly kept us ahead of the crew behind for a while; but we did not manage to catch the crew in front, who quickly caught and bumped the crew ahead of them. Stephen urged us on with even greater fanaticism, at the same time cleverly steering our boat so that the one behind could not bump us. Then suddenly, they stopped. They had been caught from behind.

A triumphant note entered Stephen's exhortations. He knew we were clear and were now not going to get bumped. The rest of us slowly realized what this meant; we had to row the whole length of the course. There would be no effort-relieving bump to allow us to pull over to the side and stop. Inevitably we began to ease up, thinking of the long distance still ahead. But Stephen was having none of it. With determination he made us keep going until we collapsed, exhausted, over the finishing line. This ensured that we would start in exactly the same position on the river the next day, with the prospect of having to do it all over again.

University College, Oxford "Rugger Eight," 1962. David Filkin is second from the left; Stephen Hawking is the cox in the distinctive boater hat and blazer.

Introduction: Finding a Boat to Steer

We obviously learned fast. To avoid the pain of rowing all the way again, we made sure we got bumped early on in the race, on each of the next three days. I remember feeling vaguely guilty about it, impressed by Stephen's dedication to our cause: but my unease soon got lost in the whirl of student life. And so did my contact with Stephen Hawking. But I did not forget the determined young man in his boater and spectacles, or his will to win through, whatever the contest.

I might easily never have met Stephen again. I joined the BBC as a graduate trainee and got immersed in television production; all I knew of Stephen was that he had gone to Cambridge to study for a Ph.D. in theoretical physics. I began to pick up bits and pieces of news from time to time, including the discovery that he was suffering from a genetic disorder, Motor Neuron disease. He would have known very quickly that he was destined progressively to lose control over his muscles: a depressing prognosis, hard for anyone to come to terms with. But Stephen's will to win and determination were to see him through this terrible blow. He had apparently been heard to say that in a way his illness had been a blessing: it had made him focus his attention on what he could do with his life. He realized his challenges would have to be all mental rather than physical from now on.

Most people are familiar with the extraordinary story of how Stephen made light of the devastating impact of his disease. He was forced to spend his life in a wheelchair; and, having had a tracheotomy, he lost the use of his voice. Yet his mind remains as razor sharp as ever. With the help of a computer and a voice synthesizer strapped to his wheelchair, he is able to continue his academic work. With tiny movements of his finger and thumb on a pressure pad, he can move a pointer on his computer screen to select common words, even whole phrases, from a specially built database; if need be, he can also spell words out letter by letter. In this way he builds up everything from drily witty jokes to the text of lectures or even whole books in his computer; and, when he wants to speak, he can trigger the voice synthesizer to announce whatever he has written.

You might imagine that this would lead to rather characterless speech. In fact Stephen has learned to manage the computer's unemotional monotone extremely cleverly to communicate his personality very clearly. To save time he tends to speak in quite brief sentences, very much to the point. At first it may seem as if he is impatient or uninterested; but very quickly, with a charming phrase here, an elegant courteous thought there, you realize it is just the intelligent efficiency of his thinking which gets to the essential point so swiftly. And above all there is his acute sense of humor. I remember Stephen being introduced to a distinguished audience at the Massachusetts Institute of Technology once. The Dean had waxed lyrical to the assembled academics, reminding them of all he has achieved. On cue, Stephen entered and neatly maneuvered his wheelchair into position while the audience rose and gave him a prolonged standing ovation. Eventually the applause died down, and there was a pregnant pause as they waited for Stephen's first pearls of wisdom to emerge. With his usual immaculate timing, Stephen turned on his American-built voice synthesizer. In fewer than ten words he won the hearts as well as the superintelligent minds of his audience. "Good morning," he said. "I hope you like my American accent."

It is all too easy to let his determined disregard for his physical limitations dominate your immediate admiration for Stephen. His academic achievements alone make him unquestionably distinguished, irrespective of any disability he happens to have. As the Lucasian Professor of Mathematics at Cambridge University, he follows in a line of outstandingly brilliant minds, at least two of whom—Sir Isaac Newton and Paul Dirac—feature in this book, as in any history of physics and cosmology. Like them, Stephen has made contributions to science which guarantee him a place in history. And, unlike many who achieve distinction in a field that is notoriously difficult for laymen to understand, Stephen has been determined to make cosmology accessible to a wider audience. He resolved to write *A Brief History of Time* without resorting to the complex specialist language of mathematics

Stephen Hawking was elected to a Fellowship at Gonville and Caius College, Cambridge. An early version of his voice synthesizer is attached to the front of his wheelchair.

usually considered so central to the study of the universe. In his introduction he explains that he was told that he would lose half his readers every time he mentioned a mathematical equation. He therefore only permitted himself one equation (Einstein's $E = mc^2$); and expressed his concern, somewhat tongue in cheek, that this might have cut sales of the book by 50 percent.

He need not have worried. The book was a sensational success. It was in the bestseller lists worldwide for years. Two or three television documentaries were made about its fascinating author; but none of them really got to grips with the science itself. As Head of Science and Features at BBC Television at that time, I was looking for new projects to launch. And so it was that I decided to meet Stephen again, some 30 years after we had rowed together at Oxford.

I was a little apprehensive when I arrived at the rather unspectacular building in Cambridge which houses his department. The last time I had seen Stephen, after all, he had been fully able-bodied; and I did not know quite what to expect. I need not have worried. He immediately put me at my ease as I sat on his right, looking over his shoulder at the computer screen while he typed in what he wanted to say. With only two of us conversing, it was quicker than waiting for the voice synthesizer to repeat what he had written. Sometimes I got the drift of what was coming before he had finished typing all the words in his sentence; and I could not resist reacting or jumping in when I thought we had agreed something. Stephen put up with this impatience on my part remarkably well. He did have one trick up his sleeve, however, to keep my impetuous instincts in order. When he wanted to complete his sentence before I got my reply in, he simply activated the voice synthesizer to say "yes" in a swift authoritative way. It was uncannily effective.

By the end of this first business meeting, Stephen had established that he was not interested in doing any more television programs about "a brilliant brain trapped in a crippled body" and was only interested if the science was the central theme. I told him I agreed; and said that I thought we would need not one but probably six separate documentaries to explain all the complexities of the universe slowly and clearly enough. But how should these programs be shaped and presented? I had come armed with the idea of using an "Alice in Wonderland" figure who would travel from one cosmological guru to the next asking all the basic fundamental questions. "Alice" would speak for the layman. Stephen was not so sure about this idea; he wanted to use a historical character like Galileo as the central inquirer. Stephen argued that Galileo's knowledge of the universe more or less equated to the lay view of today; the programs would show Galileo putting modern ideas to the test on behalf of the lay audience. We agreed that I would go away and draft six outline programs for Stephen to look at.

I quickly became deeply absorbed in the challenge. Writing drafts for Stephen was infinitely more pleasurable than taking difficult administrative decisions in the vast corporation that is the BBC. I soon realized that I was eager to get out of management and back to production again. More specifically, I did not want to hand this project over for someone else to take on the creative challenge. This was something I wanted to steer myself.

The attraction of trying to answer deceptively simple but immense and fundamental questions proved irresistible, questions like: Why are we here? What is the nature of the universe? How did it all begin? Where will it end? Discovering the answer to one question immediately gave rise to several others; and the answers were often surprising, if not beyond belief. And, instead of getting easier, it seemed to get ever more complicated. To understand black holes and time warps, you had to understand gravity and quantum mechanics, which were just as difficult to explain. It was like peeling an onion. Every time you uncovered one truth, you had to explain another three. There seemed to be no end to the peeling back of layers.

Eventually it dawned on me. I had to start from the middle of the onion, rather than the outside. If you knew nothing at all about the universe, how would you begin discovering its secrets? Our knowledge, after all, must have begun somewhere and grown, step by step, into the intensely mathematical study of cosmology that we have today. If a layman like myself could retrace those steps one by one, surely everything would start to fall into place.

The more I looked at the subject this way, the clearer it all became. I now knew what I had to do. Once Stephen had agreed and the BBC had given its blessing to my early retirement in order to make the programs and write the book, everything was in place. We had a course to follow, and a boat to row in. Once again, with Stephen's expert touch guiding the rudder, I was ready to set off up river as a complete novice. Only this time I was resolved to get to the end of the course, however long it took and however exhausting the race.

chapter one

~~~~~~

Sun, Sky and Inspiration

STEPHEN HAWKING'S MODEL OF THE UNIVERSE, AS HE WOULD READILY admit, is not something he invented on his own. It is, at present, the last link in a long chain of patient scientific observations and experiments which have steadily built up a picture of everything that has to be explained in any valid description of the universe. This is something with which brilliant minds have wrestled over the centuries; and time and time again they have discovered that every truth they unravel from improbable evidence

----

Nature had spectacular ways to persuade our ancestors that the universe was controlled by something beyond their understanding. This lightning is a thousand times more powerful than the strongest electric current we use in our homes.

reveals even greater problems than they were wrestling with before. This true-life detective story leads, clue by clue, to a universe which starts with a Big Bang, might end with a Big Crunch, and includes black holes and white dwarfs, worm holes, and WIMPs and MACHOs. It is so extraordinary that no fiction writer would have dared propose it. To understand Stephen Hawking's universe, we need to go back to the very beginnings of cosmology, long before he himself joined the list of skilled artists who have added their brush strokes to the emerging picture.

Turtles, turtles all the way down

Who first started sketching it out, and where? Some say it was someone Chinese; others make claims for one of the Babylonians, whose descendants now live in Iraq. No one really knows who was the first person to offer a scientific explanation of the nature of the universe. In any case it depends what you mean by scientific. Apparently the Babylonians thought of the universe as a huge mountain rising out of the sea with a domed sky overhead. They thought the sun entered each day by one door and left by another. They depicted stars on stones and used them for astrological predictions—not really the kind of approach which today's scientists have much time for. But why should one picture of the universe be easier to believe in than another? How can we know for sure that we are understanding the true nature of the universe, and not merely imagining what it might be like?

For a long time, people seem to have accepted that such questions were beyond their comprehension. It is easy to imagine the reactions of early cave-dwellers, confronted by the awesome fury of thunder and lightning at the heart of a raging storm, or threatened by the anger of huge waves crashing on the shores of seemingly boundless oceans. They would surely be convinced that it was not their place to know why; clearly some mighty power was telling them to stand aside and seek secure shelter wherever the Earth would protect them. Certainly, by the time early civilizations left any record of their daily lives and beliefs, there was nearly always a place for a number of gods and goddesses who between them controlled the heavens, the oceans and the Earth. Even today, we still speak of Mother Earth nurturing and feeding us as we celebrate our annual harvests.

So it would have been natural for the workings of the universe to remain firmly in the realm of religion. Any explanation of the way it all evolved had to allow for man's place in the order of things, and had to acknowledge—but not dare to challenge—the role of the gods. Within these constraints, explanations began to emerge. Over the centuries there have been some highly imaginative ideas, often beautiful and elaborate, but rarely based on anything you could call a scientific approach. In *A Brief History of Time* Stephen Hawking recalls one such model, where the Earth is a

The destructive power of a tornado still serves to remind us that forces are at work in the universe which effortlessly dwarf anything we can create for ourselves.

flat plate supported on the back of a turtle; and the turtle is the top one in an infinite tower of turtles. Unless you happen to belong to the religious sect which gave rise to this vivid image, probably with its origins in India or somewhere nearby in the Far East, you might well wonder how anyone could arrive at such an idea. But is it really any less credible than the picture of a universe exploding from nothing at all, and expanding into billions of swirling carousels, each of which includes millions of explosive fireballs of energy, around one of which our Earth is one of nine orbiting spheres? Because that is pretty much what all the calculations of modern science are asking you to accept today.

If your decision as to which picture you prefer is not to be simply a matter of faith then of course you will need evidence. And science seeks to separate fact from fiction by finding evidence. The principle is simple, but in cosmology, at least, the practice is much harder. For an idea or theory to be accepted as scientifically proven, it has to be tested in such a way that it can be tested over and over again, and the result must always confirm the theory. If, for example, you believe that pressure and temperature together determine when water turns into steam, you have to do enough experiments to show that, whenever you vary one, the other changes as well; and that the relationship between pressure and temperature always remains the same.

Unfortunately experiments to test the validity of ideas about the universe are not so easy to devise. The plate on the back of the turtle might be so large that no one, in a single lifetime, could possibly journey to its edge from the small area in the middle where people live. And, even if a determined scientific observer could get there, would he or she be able to see, by peering over the edge, if there was one turtle below, let alone an infinite tower of them? Thought of in this way, it is not difficult to see how effortlessly the imagination can come up with ideas for something as vast as the universe which leave science stranded.

As day follows night

The enormity of the problem is all the more reason to admire the ingenuity and inspiration of those who can rightly claim to be the earliest true cosmologists. They did not have to abandon belief in their gods in order to start explaining the way the universe worked. They simply had to make careful deductions from the patterns they observed. Day always follows night; the sun takes over from the moon and the stars. And, by and large, the stars appear in the same place each night. The ancient Greeks, as sailors, soon learned to rely on the positions of the stars to guide their ships. They knew that, whatever terrifying waves and storms the gods might create from time to time, for some reason they allowed the mechanics of the universe to function in a predictable, regular fashion. Furthermore, driven by their insatiably curious minds, the Greeks used their acute observational skills to try to discover precisely how these mechanics worked. Stephen Hawking cites Aristotle as one of the earliest cosmologists; but his was only one of many schools of philosophy in the ancient Greek world trying to unravel the secrets of the universe.

Of course they made mistakes—usually very understandable ones. Nothing could have felt more firm and secure than the Earth under their feet. To anyone who did not know otherwise, it would have seemed steady and unmoving. And so the Babylonian idea of a steady Earth, with the sky as a moving domed roof, made sense not only to the Greeks but to pretty well everybody for another 3,000 years. Aristotle embraced the idea of a stationary Earth as readily as anyone else; he also imagined that, for mystical reasons, circular motion was the most perfect; and that the Earth was at the center, with everything else revolving around it. But he could not come up with any way of testing his ideas scientifically. It was left to others to gather scientific evidence for the nature of the universe from the little they could observe with the naked eye.

Sun, Sky and Inspiration

Sticks and shadows

For a start, it is claimed, the Greeks noticed that the night sky, as seen at the same time from Samos and from Alexandria, had the same stars in the same patterns as each other but in different positions in the sky. Quite how the two pictures of the sky from places so far apart were brought together is not clear; but the Greeks were great sailors, and routinely traveled huge distances. It is also claimed that the Greek mathematician Eratosthenes noticed that a stick, when planted in the ground, would cast different length shadows at different times of the day. With the sun directly over-head, of course, a stick cast virtually no shadow at all. And at dawn or dusk, the shadow would be at its longest, when the sun was low in the sky, only just peeping over the horizon, to one or other side of the stick. Eratos-thenes is also credited with having somehow observed the shadows of sticks of the same length, one in Aswan and one in Alexandria, at the same time of day. Perhaps he got a friend to help him; perhaps he traveled be-tween the two places himself, and made observations at the same times on separate days. Exactly how he did it does not really matter. But what he no-ticed was that the two shadows, at the same time of day, were of different lengths.

At this point Eratosthenes had an extraordinary insight. Each stick had been placed carefully upright, at right angles to the Earth; so, if the Earth was flat, he reasoned, the two sticks would be parallel to each other. And, since the sun is so far away that the rays of light traveling to each stick were also virtually parallel, you would expect, if the Earth was flat, both shadows to be the same length at any one time of day.

This was a simple application of Euclid's mathematics; elementary geometry. On a flat Earth, when the sun was directly over one stick, casting no shadow, you would expect it to be directly over the other stick, and so equally casting no shadow. But Eratosthenes's observations showed that when the sun was directly over one stick, casting no shadow, it was casting a

Modern telescopes and photographic techniques reveal a lot more detail; but essentially it was from seeing stars like these with the naked eye that the Greeks began to calculate the nature of the universe.

---

clear shadow of the other stick. His carefully recorded data showed that the shadows from the two sticks were always a different length at any given time of day. There could be only one explanation. Even though the two sticks were both at right angles to the Earth's surface, they were not in fact parallel to each other. And that could only happen if the Earth's surface was curved.

To be fair, this would not have been such a startling revelation. The

Sun, Sky and Inspiration

For a long time models of the universe assumed the Earth was at the center of everything, and the sky was a lid with holes in it. Light from fires blazing beyond them would shine through the holes and reach Earth as starlight.

Greeks had long suspected that the Earth's surface must be curved because of their familiarity with boats. Ships were seen to come up over a distant horizon; how could that happen if the Earth's surface were not curved?

Perhaps the most significant point was that Eratosthenes had used mathematics and reasoning to arrive at his insight. He had combined careful observation and disciplined theoretical thinking, establishing a basic method with which a scientific study of the universe could be carried out. Indeed, it is the method we still use today. But that was not what immediately appealed to the Greeks. They held deep religious convictions about the significance of circles and spheres. A theory that the Earth was spherical was a theory suggesting that it was perfect; and this is exactly what Eratosthenes's experiment with the sticks seemed to confirm. In addition, his theory explained the different appearance of the night sky over Samos and Alexandria. It also confirmed Aristotle's observations of lunar eclipses (he had argued that if the Earth was any curved shape other than spherical it would not always cast a round shadow on the moon). Furthermore, when the sticks experiment and other observations were repeated, they always confirmed the same thing. This was basic science; Eratosthenes and the ancient Greeks had scientifically proved that the Earth was a sphere.

Eratosthenes did not stop there. He reasoned that you could draw an imaginary line extending each stick deep into the Earth. The point where the lines crossed had to be the center of the Earth. And, using Euclid's geometry again, he could calculate the angle between these lines. When one stick was casting no shadow, this angle had to be the same as the angle between the other stick and a line from the top of that stick to the tip of its shadow. It sounds rather complicated written down; but it is easy to see from a simple diagram.

Apparently Eratosthenes also knew the distance from Aswan to Alexandria (in other words, the distance between his two sticks). So he knew the length of the small section of the circle subtended by the angle at the centre of the earth which he had calculated from the one stick and its shadow. And from knowing this distance, and the angle which subtended it, he could calculate the length of the rest of the circle running all around the Earth. He had now not only proved the Earth to be a sphere; he had also come up with a way to measure its circumference. And his calculations were remarkably close to the figures we can calculate today.

The harmony of the spheres

This was an immense step forward. Finding a way to calculate the circumference of the whole Earth meant that mathematics could take our understanding of the nature of the universe far beyond what could be gleaned from observation alone. For a while, led by Pythagoras, Greek philosophy got carried away with the wonder of mathematics. The first calculations were made to measure the distance from the Earth to the moon and the sun (sadly, the results were quite wide of the true figures, but the methods used were improved on, and were to last for centuries). The Greeks' mathematical principles were sound enough; it was their lack of accuracy in measurement which let them down.

Pythagoras saw the mathematics in music, and envisaged everything being described in terms of mathematical formulas. He proposed a universal theory, "the harmony of the spheres," which used the scientific precision of mathematics to support the traditional Greek belief in the perfection of spheres and circles. However, as rather too many attempts at

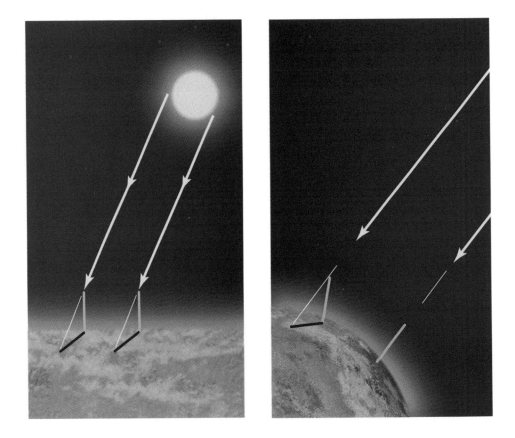

harmonious mathematical descriptions ended up in ugly imperfect numbers (and not the simple basic ones Pythagoras had imagined would be at the root of everything) this ideal became somewhat tarnished. Nevertheless, the introduction of mathematics had finally opened the way for man to explore the universe scientifically, to go beyond what could be seen with the naked eye.

Yet there were some observable facts which did not easily fit into a mathematical explanation. One of these was a flaw in the perfect composi-

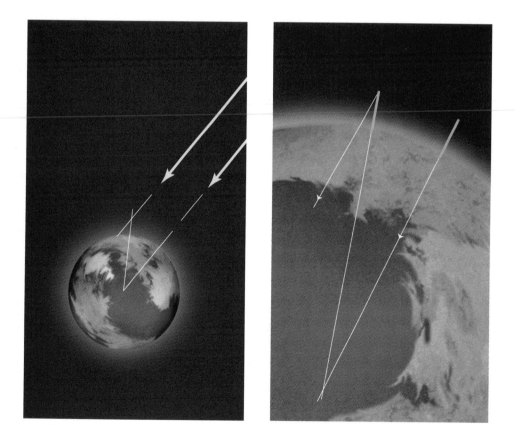

tion of the night sky. The Greek astronomer Hipparchus pointed out that not all the stars that shone seemed to keep the same regular position as the majority did. These "wandering stars" appeared to move off in one direction and then double back on themselves. They would also appear to get brighter and dimmer from time to time. Was this just another example of mathematical irregularity to discredit Pythagoras's vision of perfect harmony?

A clue to what he had seen remains in our description of the universe today. The Greek word for "wanderer" (*planetos*) gave rise to the name we still give to the "stars" Hipparchus observed. They turned out not to be stars but "planets": wanderers in the night sky whose movements had to be explained in the context of the emerging picture of the universe. At the time, five of them were visible to the naked eye: Mercury, Venus, Mars, Jupiter and Saturn. For the Greeks, our Earth may have been spherical, but it felt solid and stationary as they stood on it and observed the movements in the heavens. Was there a clear mathematical pattern to the way the sun and the night sky, wanderers and all, were seen to be moving around us?

Painstaking observation over many nights enabled the Greeks to plot the paths the planets took. They soon realized that many pieces of the paths were curved; could they be segments of circles? Plato appealed to all the Greek schools of philosophy to turn their minds to working out how a system of perfect circles could explain the erratic paths of the "wanderers." If the planets, along with the moon and the sun, encircled the Earth, then symmetry would triumph and mathematical perfection in the universe could be maintained. It would be exactly as Aristotle had imagined.

The Greeks had no idea of what today's technology would reveal. But they saw enough to be convinced that the workings of the universe depend on the unsurpassable beauty of perfect spheres and circles.

Orbits within orbits

A number of ingenious proposals emerged. One approach suggested that, if the Earth was not in fact at the precise center of a planet's orbit, then at one point of the cycle it would be much closer to the planet than any other; and at the opposite point it would be as far away from it as it ever could be. This would explain the change in brightness. But it could not account for the changes in direction of the planets' movements across the sky. Another approach suggested that, while a planet might follow an overall circular path, with the Earth at the dead center of the orbit, there could be other tinier circular movements going on. They were called epicycles; and they involved the planet having a small orbit around a central point which itself orbited the Earth in a perfect circle. This would again allow the planet to be at different distances from the Earth, to explain changes in its brightness; in addition, it would allow the planets to double back on themselves. But this model was flawed too; none of the observed planetary orbits would fit into a configuration like this closely enough for the theory to be convincing.

Finally, in the second century A.D., an ingenious astronomer named Ptolemy built a complex model which combined both these ideas. The planets did move in epicycles,

Medieval scholars nearly always portrayed Ptolemy's model of the universe as a series of simple circular orbits with the Earth at the center.

he argued; but the Earth was not precisely at the center of each planet's overall orbit. Following this principle, Ptolemy drew up models for the orbits of the five known planets, and the sun and the moon, around the Earth. In addition, surrounding them all like a shell, was an outer sphere, the sphere of the fixed stars. He put the Earth at various different distances from the centre of the seven orbits; and added a number of epicycles into each one. In this way the changes in brightness, and all the irregular movements of the planets, could be recreated pretty much as they were seen from Earth. At last, it seemed, Aristotle, Pythagoras and Plato had achieved what they had all wanted. The whole universe could be accurately described in terms of spheres and circles.

What Ptolemy, and everybody else for some time, failed to realize was that you could make any orbit fit into this kind of pattern, irrespective of its true shape. If a planet takes longer to complete its orbit than a perfect circle will allow, it is easy enough to tuck away the extra time in a number of little circular back-loops all along its path. You just adjust the size of the loops to match the time you need to allow for. But so convincing was the idea of the Earth being the stable center, with everything else moving around it, that this basic picture of the universe was to last for centuries.

There was at least one major problem with Ptolemy's model. In

But Ptolemy himself could only make the sun, the moon and the planets circle the Earth by adding complicated epicycles to these basic orbits.

Sun, Sky and Inspiration

order to make the path of the moon fit in with the overall picture, it had to be twice as close to the Earth at some times than at others. And that would imply that the moon would look twice as large on some days as on others. This problem cast doubt on the truth of the Ptolemaic model. But it was near enough at the time to the observed reality to be convincing. It was perhaps not the first scientific model of the universe, but it was the foundation

Early models of the universe were often portrayed in pictures which owed as much to art as to science. But they vividly reflect the excitement with which scholars pursued their conviction that the whole universe could be scientifically explained.

on which our future understanding of the universe was to be built. And it had the supreme advantage that it hardly offended anyone. Most religions could find a place for a universe of spheres within spheres in the wider context of their faith. If anything, this strengthened the religion's appeal by showing that it embraced what science had established as its current version of truth. If the religion's god or gods had created the universe, this scientific model simply described the nature and workings of that universe.

The first rung of the ladder

The fact that the model eventually turned out to be wrong was no embarrassment. After all, science has to set up hypotheses and test them within the constraints of the times. Its discoveries are only valid for as long as experiment and observation continue to support the hypotheses as true. Science cannot promise eternal truths; only the elimination of false hypotheses and the establishment of what is currently the most likely explanation of an aspect of reality. Whatever the shortcomings of some of their theories, the Greeks succeeded in placing mankind on the first rung of the ladder that would eventually lead us to the way we perceive the universe today. Without them, Stephen Hawking might not have been inspired to develop his own model of the universe. Ptolemy, and Hipparchus and Eratosthenes before him, created a scientific picture of the universe which was to last far longer than any other. And it was based only on the observations of the naked eye, a few experiments with sticks and shadows, and the fathomless brilliance of the human mind.

~~~~~~~~~~

Christian
Complications

PTOLEMY'S MODEL OF THE UNIVERSE TURNED OUT TO BE AS ACCEPTABLE
to the Christian Church as it had been to the religions of the ancient world.
Following the crucifixion of Jesus, Christianity had rapidly eclipsed other
belief systems throughout Europe. Its scriptures very clearly preached the
creation of the world by God, with man and woman—in the form of Adam

―――――――――――――――

Did God create the Earth at the heart of the universe? Or are we
only, as this artist suggests, a sideshow in the theater of the
cosmos viewed from an alien land? For the early Christian
Church there were no doubts. Man had to be at the center of
everything.

and Eve—on Earth at the center of everything. And this was totally consistent with the Earth-centered universe described by Ptolemy.

Increasingly, the Church became the exclusive patron of scholarship. Literacy was required for the study of the scriptures, and only the Church could afford to teach people to read. Eventually all men of learning had to have the patronage of the Church if they wanted to study, and at the same time feed themselves and keep a roof over their heads. This meant that scientists were also priests or monks, dedicated both to the study of science and the spreading of the Church's teaching. They unhesitatingly taught the story of God's creation of a Ptolemaic universe. Science and religion were as one.

The Copernican revolution

It is not surprising, therefore, that Ptolemy's model of the universe went unchallenged until the sixteenth century. To begin with, it suited the Church's teachings very well; and, in any case, there were no new technologies available to improve observation. However there was some intelligent thinking going on. Nicholas Copernicus, a Polish priest, was becoming increasingly convinced that the planetary orbits which Ptolemy had proposed needed far too many adjustments with circular back-loops, or epicycles. He realized that he could eliminate many of them if he made the sun the center of everything instead of the Earth. He was fully aware that this would seem like heresy to the Church, with its conviction that God had created man on Earth as the center of the universe. So it must have taken considerable courage for him to publish his ideas in 1543. Indeed it seems that, before doing so, he tested the waters by letting the model he proposed be circulated anonymously; only when it was not immediately condemned did he acknowledge it as his own. By then he was on his death bed, and even at that time it was apparently only the pleading of his amanuensis, Rheticus, which per-

Nicholas Copernicus (1473–1543) offered a revolutionary alternative to Ptolemy's model of the universe. He assumed the sun was at the center of the universe, not the Earth.

suaded Copernicus to publish.

The rather surprising lack of reaction from the Church was probably simply because it did not take Copernicus seriously. While his model certainly simplified the Ptolemaic model, it had the serious drawback that, even if the planets orbited the sun, their observed movements did not quite match the circular orbits Copernicus proposed. He could only improve on Ptolemy's circular loops with a flawed alternative. It probably did not seem much of an improvement as far as the Church was concerned; certainly not enough to threaten the established order of hundreds of years. But, if he failed to concern the Church, Copernicus certainly succeeded in arousing the interest of other scientists.

The thinker and the information-gatherer

One of these was Johannes Kepler, a German astronomer who eventually lived in Prague. He had little opportunity for observational work; but he was a brilliant theoretical thinker. He was trying to work out why any heav-

SIGNORVM STELLARVQVE DESCRIPTIO CANONICA
ET PRIMO QVAE SVNT SEPTEMTRIONALIS PLAGAE

FORMAE STELLARVM	Longitud.		Latitudinis		M
VRSAE MINORIS SIVE CYNOSVRAE	pars	sc	pars	sc	
In extremo caudae	53	30	66	0	
Sequens in cauda	55	40	70	0	4
In eductione caudae	69	20	74	0	4
In latere quadranguli precedente australior	83	0	74	20	4
eiusdem lateris borea	87	0	77	40	4
earum q in latere sequitur australior	100	30	72	40	2
Eiusdem lateris borea	109	30	74	50	
Stellae septem quae sede magnitudis tertia					
Et quae circa Cynosuram informis in latere sequente ad rectum					
lineam maxime australis	103	20	71	10	

VRSAE MAIORIS QVAM ELICEN VOCANT					
Quae in rostro	78	40	39	50	
In binis oculis precedens	79	10	43	0	
Sequens hanc	79	40	43	0	
In fronte duarum precedens	79	30	47	10	
Sequens in fronte	81	0	47	0	
Quae in extra auricula precedente	81	30	50	30	
Duarum in collo antecedens	84	40	43	50	
Sequens	92	40	44	20	
In pectore duarum borea	94	20	44	0	
Australior	93	20	42	0	
In genu sinistro anteriori	89	0	35	0	
Duarum in pede sinistro priori borea	89	50	29	0	

enly body should orbit another in the first place; and he came up with the idea that some kind of magnetic force was at work. If this remained constant, it would at least keep one body at a regular distance from the other, thus creating circular orbits. But he was unhappy about the idea of such a force being effective over the great distances involved.

He must have become even more agitated when his next inspiration struck. He had the idea that some other shape of orbit, perhaps elliptical rather than circular, might somehow make sense of the Copernican model of the universe. In other words, if everything orbited the sun, and not the Earth, and if the orbits were always oval-shaped, then every planet in the sky could perhaps follow a straightforward path without the need for complicated epicycles and other adjustments. Our observations of their movements could be perfectly described in simple, elegant orbits. But this would make a nonsense of the more perfect circles which everyone believed were the basis of the orbits, as well as destroying his magnetic force theories. The only way to progress would be for him to gather new, more accurate observational data on the precise orbits of the planets.

He had heard of someone who he thought might be able to help—Tycho Brahe. He was considerably older than Kepler, and already renowned for his astronomical observations. He held a powerful position at the Danish court; it was the custom at that time for astronomers and mathematicians to be appointed advisers by the powerful kings of Europe. In return for astrological predictions which might help the king make political decisions, these influential courtiers were often supported in their more scientific pursuits. So it was that the King of Denmark gave Tycho Brahe an island from which to observe the heavens. He also paid for the construction of the most precise observational instruments yet made (though they were, in fact, little more than refinements of the instruments developed by the Ancient Greeks after their experiments with sticks in the ground). Kepler had heard that Brahe was methodically mapping the skies, making more accurate and more frequent observations than ever before.

There are a number of versions of what happened next. The most ap-

Left Copernicus kept painstaking records of the stars he observed, carefully noting their position and size. It was the kind of organized observation which would eventually lead Johannes Kepler (1571–1616) to find a theory which supported Copernicus's proposal that the sun, not the Earth, should be at the center of the universe.

Christian Complications

Tycho Brahe

Tycho Brahe, who, before telescopes were invented, used instruments like his sextant (*right*) to map the stars and give accurate readings using only the naked eye.

pealing one may be apocryphal (the truth has a habit of being far more prosaic than people would like). But, according to the most dramatic version of the story, Kepler decided to undertake a long and arduous trek across Europe to meet Brahe, to see if his observations would in any way support the idea of elliptical orbits.

If the story is correct, Kepler must have been in despair when he finally arrived. The Danish astronomer allegedly refused to see him, presumably fearful that Kepler would use Brahe's data to announce some great new truth about the universe and give Brahe no credit for it. According to this version of the story, Brahe was an extremely careful observer but rather uninspired at interpreting his data. He could not face the possibility that he was missing something significant which Kepler had caught on to without doing any of the observational work.

It is said that Kepler returned home with nothing to show for his long journey. Brahe kept plugging away at his observations; and still could not understand their significance. Eventually he had an idea. Perhaps if he gave Kepler the data on a single planet, he could find out what Kepler made of it and then work out how to interpret the movements of all the rest. That way Kepler might get the credit for understanding one orbit, but Brahe would be seen to have discovered how all the others worked. It is claimed that it was Brahe who took the long journey this time, in order to show Kepler the data he had collected about Mars. Kepler was naturally delighted, especially when it became clear that the orbit of Mars simply had to be elliptical. But imagine Brahe's dismay when Kepler told him that he no longer needed to see the data for the other planets. What he had seen of Mars alone was sufficient to confirm his idea that planetary orbits were not circular. They must all have elliptical orbits.

Sadly for romantics, the true story may be less absorbing. Apparently, when the King of Denmark died, his successor dismissed Brahe. Some say that the astronomer was a vain and arrogant man, difficult to get on with and disliked by everyone at court except the old king. He was able to get a new post in a different court and a different country, at Prague, and, having heard of Kepler, resolved to appoint him as his assistant there. Perhaps, as he got older, and still could not work out what his careful observations were revealing about the universe, Tycho Brahe began to fear dying without concluding his hard work. Even though, undoubtedly, Brahe and Kepler did not get on—each secretly hoping to prove their own rather different pet theories correct—Kepler did become Brahe's assistant and so they spent a considerable amount of time together. They worked together on

Christian Complications

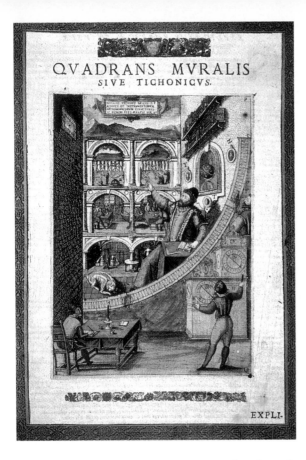

Some instruments at Brahe's island observatory were giant and expensive constructions.

Brahe's observations of Mars; but it was not until 1609, after Brahe's death, that Kepler published his idea of elliptical orbits. Even if there was a dramatic rivalry between the two men, they certainly collaborated in the end. Brahe's careful observations and Kepler's inspired intuitive theory were both equally important in the discovery that planets must have elliptical orbits—around the sun, as Copernicus had predicted, and not the Earth. Sadly, it seems, Brahe died before the true value of his observations was publicly acknowledged.

Galileo's heresy

Curiously, the Church was still not unduly concerned. Perhaps Prague and Denmark seemed too far away to worry about, when Italy's own leading authority, Galileo Galilei, was still steadily teaching his students that the universe had the Earth at its center. Galileo was Professor of Mathematics at Padua, and well known and respected for his scientific work. According to whose version you decide to believe, he was either a model scientist or a no-

Galileo Galilei (1564–1642) built several surprisingly simple telescopes all based on the same design. These (*below*) were undoubtedly his; but no one now knows if either was the one he actually used to make the historic observations which brought him into conflict with the Church.

torious plagiarist who took the credit for other people's ideas. Stephen Hawking certainly regards him as the father of modern science, at least partly because of the way he refused to teach new ideas until he was satisfied that they had been properly tested experimentally. But, once he was satisfied that science confirmed something to be true, he would stand by it, whatever difficulties this might cause him.

Whichever view you take of his work, it cannot be denied that Galileo made an immense contribution to modern science. (His death, incidentally, was 300 years to the day before Stephen Hawking was born. This is a bizarre coincidence that probably has nothing whatsoever to do with the contribu-

Christian Complications

tion each of them has made to our understanding of the universe—but fascinating nevertheless.)

Galileo did significant work in discovering the laws of motion; he calculated, for instance, what happens when a ball drops from a height to the ground below. Irrespective of the size of the balls used in the experiment, the rate at which they gather speed is always the same. Whatever the legend says, this could have been simply tested by rolling two balls of different weight and size down the same gradual slope, without resorting to dropping them from the top of the Leaning Tower of Pisa. The overall conclusion, that all things drop at the same rate of acceleration unless some other force prevents this, would still be the same. And using the slope, which Galileo did, made it much easier to time and observe the movement of the balls. In this way, Galileo took the first steps toward understanding gravity, obviously an important scientific development. But perhaps of more immediate significance were some other observations he made which finally brought the Church and science into conflict.

This came about because he was able to make use of a significant improvement in technology. He did not so much invent the astronomical telescope from scratch as have the intelligence to assemble it from what others had done elsewhere. Lenses had been in use for a while as simple magnifying devices, but the magnification available from a single lens was pretty limited. It is said that it was only by accident that two children, playing in an Amsterdam shop selling scientific instruments, discovered what happens if you look through two lenses at the same time. The owner of the shop put two lenses into a tube, one at each end, and sold it as a magnifying instrument. The Dutch scientist Leeuwenhoek used this idea as a microscope before Galileo turned it toward the sky in 1609 and used it as a telescope. And what Galileo eventually saw was sufficient to convince him that the Copernican model of the universe, with the sun at its center, had to be the true picture. The Italian professor very publicly and loudly made the startling assertion—startling at least for the Church—that Ptolemy's Earth-centered

Because Galileo was such a successful publicist, people everywhere started to look at the skies for themselves, to find out just what he had seen. *Below* With no other means to record his observations, Galileo made these careful sketches of how the moon looked through his telescope on different occasions.

model of the universe could no longer be supported by science.

What convinced him so quickly was that the telescope immediately revealed all sorts of imperfections which cast doubt on the idea that the universe was made up of perfect circles and spheres. There were blemishes on the more closely observable heavenly spheres, such as the sunspots, and craggy craters on the moon. But, through his telescope, Galileo observed two phenomena which dealt the real death blow to the Ptolemaic universe. The first was the fact that Jupiter had moons in

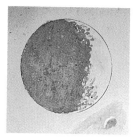  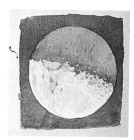

Christian Complications

In 1979, the *Voyager* space craft photographed Jupiter and its moons. Galileo identified the four largest (*from the left*) Ganymede, Europa, Io, and Callisto. All he saw with his primitive telescope were four "spots," whose positions changed from day to day.

orbit around it. (The Church had always supported the Ptolemaic universe because it allowed everything in the universe to orbit the Earth; but here were some heavenly bodies which were clearly orbiting something else.) The second phenomenon Galileo observed were shadows on the surface of Venus. By studying how these shadow patterns changed over time, it became clear that they had to be the result of Venus orbiting the sun. The case for an Earth-centered universe was collapsing. The Copernican model was clearly much more likely.

This time the Church was not as unperturbed as it had been before, for perhaps three main reasons. Firstly, it could not ignore the widespread impact of Galileo's ideas. Because Galileo published his views in Italian rather than academic Latin, the public at large began to support them, to which the more traditionalist professors responded by urging the Church to reassert the validity of the Ptolemaic view. Secondly, Galileo made things worse by suggesting that, where the Bible conflicted with common sense and science, it was being allegorical. He even went as far as to say that anyone who could not see the clear logic of the evidence against an Earth-

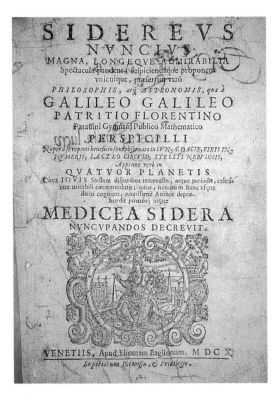

If Galileo had not published his own tabloid *Starry Messenger*, announcing his observations to the public at large, he might have escaped the full wrath of the Church.

centered universe was being stubbornly stupid. This was tantamount to blasphemy, in effect declaring that the Pope and his advisers were fools. Thirdly, the rise of Protestantism was threatening the Catholic Church. The time had come for it to assert its authority and to reaffirm its traditional world view. In 1616 the Church declared that Copernicanism was false, and commanded Galileo to abandon the doctrine.

Joseph-Nicolas Robert-Fleury (1797–1890) portrays Galileo's argument with the Church authorities in 1632, just before he was condemned by the Inquisition.

Christian Complications

Galileo probably had little choice but to acquiesce, if he wanted to continue his work as a scientist in other fields. Putting the record straight about the nature of the universe might not have seemed so fundamental that he should give up everything else for it. So he did as he was told; and in 1623 his obedience was seemingly rewarded. A new Pope was appointed, who happened to be a long-time friend of the professor's. Galileo immediately sought to have the 1616 decree revoked; but the new Pope was sensitive to Church politics and would only agree to a compromise. The Pope conceded that Galileo could publish a new account of both the traditional Ptolemaic and the Copernican models of the universe, without supporting one against the other. He also insisted that Galileo should point out that it is not, ultimately, in man's power to know how the universe works, since that would constrain the omnipotence of God. (Interestingly, this response was very similar to the one Stephen Hawking and other scientists received when they explained the latest cosmological theories to the Pope over three hundred years later.)

Galileo wrote his new book but, despite following all the conditions laid down by the Pope, it aroused even greater support for Copernicanism. Galileo simply could not—or would not—muster a convincing case for the Ptolemaic model and also be true to what his telescope had revealed. The Church finally had him brought before the Inquisition and sentenced to house arrest. He also had to renounce Copernicanism for a second time. But by then it was too late. Although Galileo complied with the Church's command, it was already widely perceived that science had demonstrated new truths about the universe which could not be erased by religious dogma.

Constraints on knowledge are notoriously difficult to maintain; and it would not be long before the Church would have to reconcile itself to the results of scientific observation. Kepler's elliptical orbits and Galileo's observations were all making sense of the Copernican universe; the only problem left was that of explaining the force which kept the planets in orbit. Once this had been done, the Church would surely have to admit defeat. It badly needed to find a way of adjusting its views without appearing to make a complete volte face.

Newton's irresistible force

Isaac Newton unwittingly provided an expedient way out for the Church. He is, of course, associated in the popular mind with apples falling from trees and the discovery of gravity; but he made immense contributions to science in other directions too. In fact he has been dubbed by some "the father of modern science" in place of Stephen Hawking's preferred candidate, Galileo. In the end, what matters is that both are of colossal importance. Effectively, Newton extended Galileo's work by explaining the direct relationship between a force and an object using a series of mathematical formulas.

Although he had had no training in mathematics at school, Newton had such an aptitude for the subject that he soon became the Lucasian Professor of Mathematics at Cambridge University—the post which Stephen Hawking now holds. At the heart of his work is the idea that any object will continue to move in a straight line at the same speed unless a force is introduced to alter this. When an object is at rest it is because some force has stopped it moving; it is friction, for example, and air resistance which eventually make a rolling ball come to rest. And when

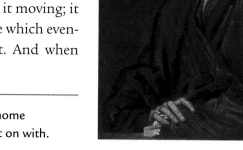

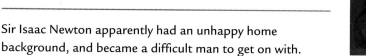
Sir Isaac Newton apparently had an unhappy home background, and became a difficult man to get on with.

Christian Complications

Falling apples may have influenced Newton but the complicated mathematics involved in his theory of gravity suggest he must have taken more than just a moment of insight to come up with the complete theory.

---

something moves at a faster or slower speed, or changes direction, it is because some force has acted on the object to produce the change.

Newton used mathematics to show that changes in speed and direction of movement were always proportionately related to the mass of the object and the power of the force involved. The mass of an object is a property related to its size and weight. To be more precise, the mass of an object can be defined by the effort or amount of force it takes to get it moving, or to increase the speed of its movement if it is already moving.

These laws of motion, which are still taught today as the fundamentals of physics, paved the way for an explanation of gravity. Newton argued that every body or object attracts every other body or object with a force called gravity. A body with a very large mass will visibly draw a body with much smaller mass toward itself. In this way the Earth draws an apple toward its surface. The apple also draws the Earth toward itself. But, because the apple has such a small mass compared with the Earth, it attracts the Earth toward it only by an imperceptible amount. Thus, it appears as if the

Newton's famous book, *Principia Mathematica*, implies in its title exactly what he had achieved. It was the first completely mathematical description of the universe which totally fitted observations up to that time.

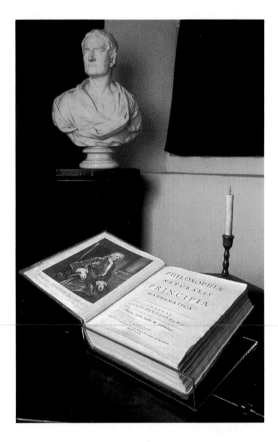

pull is all in one direction, with the apple falling to the Earth. In fact gravity is such a weak force that two objects of similar mass will also only affect each other imperceptibly. Two apples of similar mass, on a plate, will not therefore be drawn to each other, even though there will be the slightest of attractions between them. Furthermore, the greater the distance between the two objects, the weaker the attractive force of gravity becomes.

This explanation was totally consistent with the observations Galileo had made—that objects of different weight and size always fell to Earth at the same speed. The mass of the Earth is, after all, so huge in comparison with any object a man can drop in the vicinity of the Earth's surface that any difference in mass between the dropped objects will have a negligible effect on the equation. It is a bit like trying to distinguish between the impact of a gale force wind on an oak leaf and its impact on a birch leaf.

But, when other massive objects—like the sun and the moon and the planets—were considered along with the Earth, Newton realized that the gravitational attraction between them would be significant, even over great distances. With no other force to stop them, all these heavenly bodies would

Christian Complications

The constellation of Orion the Hunter is familiar to most people, but a telescope image gives much more detail than the naked eye.

be moving in some direction or another in space. Normally this movement would be a straight line, unless one heavenly body got close enough to exert a gravitational pull on another. The force might not be strong enough to pull the two bodies completely together; but it could certainly be enough to bend the line on which one of them was moving. It might even draw the less massive object into an orbit around the more massive object, if the correct balance was struck between the mass of the objects, their speed and direction of movement, and their distance apart.

Newton worked out a mathematical equation to describe this relationship, and decided to see how it would apply to the solar system. When he applied this formula for gravity to Kepler's elliptical orbits, Newton found that it fitted almost perfectly. The orbits for Mars, Jupiter and Saturn were exactly right; but it was later discovered that there was a tiny discrepancy in the orbit of Mercury. This was immediately put down to the imperfection of observations, since Newton's equations explained all the other orbits so well. The conclusion was obvious: Copernicus

Newton's tiny reflecting telescope would have shown him much of what this modern photograph reveals. He was the first person to put a mirror into a telescope, allowing pin-sharp, increased magnification within a much shorter tube. His telescope, hardly more than a hand's span in length, became the model for subsequent giants which easily dwarfed the observers looking through them.

Christian Complications

and Kepler had been right all along. And Newton had shown that it was gravity which kept the planets in elliptical orbit around the sun.

Newton's laws, formulated over three hundred years ago, have proved so accurate that they are still used to predict the speed and trajectory needed to put satellites in orbit around the Earth. His was a colossal intellectual achievement; and the Church could no longer resist the evidence that the sun, not the Earth, was at the center of the universe. What indirectly made this easier for them to accept was something else which Newton believed. He was not happy with the idea of there being an outer limit to the universe—in Ptolemy's model the sphere of the fixed stars. Newton had helped develop newer, more powerful telescopes, which had shown that, in fact, not all the stars were fixed; careful observations showed that they moved, if only by almost imperceptible amounts.

As Newton's laws of motion were based on the idea that nothing is naturally at rest, he reasoned that all heavenly bodies should behave like the more easily observable nearby planets. They would be constantly on the move, with gravity dictating their paths as they came into each other's gravitational fields. And if everything was always restless like this, where and how could the limits of the universe be drawn? There was no logical requirement for there to be an edge of the universe. This led him to suppose that there were no limits to the universe. It could be infinite in both space and time.

Although, in one sense, this contradicted Church doctrine (in that it made it difficult to specify a time and place for a moment of creation), it did at least embrace the idea of the infinite and the eternal. This was very much how the Church saw God. He would, in His infinite power and wisdom, have created the universe to be as infinite and eternal as Himself; how He did so would be beyond the comprehension of mortal men with their finite lives.

For the moment science and religion were uneasily reconciled; but neither had emerged totally unscathed from the intellectual conflict. Religion had had to concede that science was right in its discovery that the Earth was just one of the planets orbiting the sun, and was not the center of everything. And the Church could not pinpoint the moment of creation it had always preached. But science was no better able to suggest why the universe was there. Nor could it solve a crucial problem with Newton's picture of an infinite universe under the influence of gravity. If every object exerted an attractive force on every other object, why had all the stars in the universe stayed apart from each other for so long? In an infinite and eternal universe, surely everything would eventually be pulled by gravity into a single huge conglomerate. And this did not seem to fit with the universe that had been regularly observed for thousands of years.

Nevertheless, Newton's mathematical laws of motion and gravity so neatly explained those observations that his model of an infinite, eternal universe soon became as widely accepted as Ptolemy's model had once been. But, unlike Ptolemy's Earth-centered version, the Newtonian infinite model was only to remain unchallenged for just over two hundred years.

chapter
three

~~~~~~~~~

Seeing the Light

Seeing stars

IN THE WAKE OF NEWTON'S IDEA OF AN INFINITE AND ETERNAL UNIVERSE came a whole new enthusiasm for staring into the sky. With the help of increasingly powerful telescopes, people became fascinated with finding out

---

Newton thought that there was no reason to set a limit to the extent of the universe; stars like these in the constellations of Taurus (*center*) and Pleiades (*the cluster above*) could stretch out beyond us forever. But would there be a limit to how far we could see?

how far into the infinite they could peer and speculating on what they might find. By inserting mirrors into the telescope, Newton had found a way to produce sharp images which greatly magnified everything Galileo had been able to see. But this improved technology did not at first lead to any great new insights. While the planets and comets provided exciting new quarry for astronomers to track down, beyond them all that could be seen were familiar points of light: nothing but more and more stars. The nearest stars, under closer scrutiny, were just as the earliest observers had seen them with the naked eye—sources of bright shining light. The latest telescopes showed only what had already been observed; there were just more stars, magnified to look bigger.

So, instead of discovering new varieties of heavenly bodies deep in the universe, astronomers found themselves speculating on what the growing number of star sightings might mean. In about 1750, Thomas Wright found that more stars were to be seen clustered together when looking in one direction than in others. The idea grew that we were in some kind of family of stars; and that there might well be other families or clusters like it. The German philosopher Immanuel Kant observed "dust clouds," or nebulae, and suggested that these could be distant clusters of stars like our own; our clus-

When more powerful telescopes looked beyond the nearest stars, all they appeared to see at first were more stars. But some of the points of light now appeared to be not the tidy pinpoint light of a star, but a more irregular blob of light.

Thomas Wright (1711–1786) drew these imaginary clusters of stars to explain how he thought they might group together to form what looked like a single blob of light.

ter would be one galaxy and these nebulae would be different clusters, or galaxies, as we now call them. Meanwhile, the French astronomer Charles Messier began to map these clouds, trying to establish a pattern for the universe.

In 1785, William Herschel and his sister Caroline finally found a new heavenly body when they discovered the planet Uranus; they also spotted within certain nebulae pinpoints of light, which they announced were separate stars in actual clouds of dust and gas. Then William Parsons, the third Earl of Rosse, built his Leviathan telescope in 1840 at Birr Castle in the middle of Ireland. This enormous 10-ton tube was at the time the most powerful telescope in the world; and it was finally confirmed that there were individual stars within galaxies. What had been seen with smaller telescopes as rather uneven blotches of light, and so assumed to be irregular-shaped stars, or stars surrounded by clouds of dust, were now seen to be groups of perfectly formed stars whose individual light had been lost in a blur of brightness from the whole galaxy.

All the colors of the rainbow

Despite the new knowledge they were acquiring, astronomers were still frustrated. If all that could be seen beyond our immediate solar system were clusters of light, it seemed unlikely that we would be able to learn much more about the universe from observation alone. Perhaps we had reached the limits of human knowledge. It was a sobering and frightening thought.

However, as it happened, other scientists were already making progress without realizing how relevant their work would be to cosmology. In 1816 Joseph von Fraunhofer, a German lens manufacturer, was testing the glass he used in his lenses at his laboratories near Munich. He had seen something unusual using artificial light, and he wanted to check if the same

With his huge Leviathan telescope at Birr Castle in Ireland, William Parsons was the first person ever to see a spiral shape in the sky.

Top Parsons sketched the whirlpool galaxy but did not know at the time what it was.
Bottom Recent photographs confirm how accurately Parsons drew what he had seen.

thing could be seen in the full spectrum of colors refracted from sunlight. Refracting light into its spectrum of colors was something Newton had demonstrated around a hundred years earlier. It is one of the things most schoolchildren are still taught today in their early physics classes, using a triangular prism of glass to break up a beam of light into its separate wavelengths. The result is a glorious range of colors, from red and orange at one end, through yellow, green and blue, to indigo and violet at the other. Joseph von Fraunhofer simply wanted to see if any imperfections in his lens glass would be revealed in the rainbow pattern produced by the refraction. At first he had been using artificial light (produced by heating sodium until it gave off a yellowish light) to test his glass. And he had noticed that the light refracted from this lamp had one or two mysterious gaps in it—dark lines where the continuous spread of the colors was interrupted at very precise positions every time he refracted the lamp light. But the sodium light revealed only part of the spectrum, and von Fraunhofer wanted to examine the whole spectrum to see if the lines were apparent in sunlight too. In the precise conditions under which he was testing his lenses, he not only saw the rainbow effect produced by refracting the light, he also noticed a large number of quite clear lines all across the spectrum; there were some distinctive dark ones and some less clearly visible light ones. They are not so easily spotted in a school laboratory, but under von Fraunhofer's conditions they could definitely be seen.

He soon tried heating chemicals other than sodium, and refracting the light they produced. Again, the lines or gaps were there, but now they were in different positions. He did not know what caused them but, for each element, the pattern of lines was different from the pattern produced by any other chemical used to make light. They were a bit like the bar codes used in supermarkets nowadays to encode prices and descriptions on goods; each pattern of light and dark lines, seen embedded in the color spectrum, was a kind of light fingerprint by which, if von Fraunhofer had only realized it, the chemical element being heated could be identified. For the moment, however, all he knew was that he had seen these lines and that he should publicize their existence, in the interests of science.

We now know that these lines are, in fact, points in the spectrum, or

In this spectrum the Fraunhofer lines can clearly be seen; their pattern reveals the precise chemistry of the light source.

specific wavelengths of light, where each element either absorbs light—creating an absence of light or a dark line, or radiates light—creating a brighter color or extra light. It has to do with the subatomic structure of each element and the way it responds to the input of energy. All of which is not immediately important for understanding the significance of von Fraunhofer's discovery. What matters is that there are chemical fingerprints in each sample of light which can be seen by refracting that light. And by identifying each of these "fingerprints," it is possible to tell which chemical elements are present at the source of the light.

Von Fraunhofer was not the only one who did not immediately understand the significance of his discovery; in fact it was not until around 1880 that William Huggins discovered that these Fraunhofer lines were the fingerprints of the elements. What is more, he realized that they were going

Seeing the Light

to be useful in telling us what the sun and the stars were made of. When he refracted the light from the sun and compared it with the light from a star, not only did he see that each gave off light with an identical fingerprint, he was also able to distinguish in both samples the clear fingerprints, superimposed on each other, of helium and hydrogen. The inevitable conclusion was that the stars and the sun were similarly made up of hydrogen and helium, burning or reacting in some way, like a giant version of one of von Fraunhofer's lamps, to give off heat and light.

This was, in itself, significant science. But, philosophically, the most important thing he had established was that the sun and stars are no different from each other. In other words he had discovered something much more humbling to mankind than even Galileo's observations that the Earth was not at the center of everything. The sun, at the heart of our solar system, was far from unique. It was just another star, one of billions, all made up of the elements hydrogen and helium in some way combining to give off heat and light all across the universe. It seemed that our place in an infinite universe was both arbitrary and insignificant.

This time neither the Catholic Church nor any other branch of the Christian religion wanted to challenge the scientific view. If it diminished man even further in comparison to God, this served only to underline God's awesome power and infinite wisdom. For many scientists, already uncertain of the validity of any religious faith, Huggins's discovery suggested that science was indeed independent of religion, and that a complete understanding of how the universe worked might eventually be arrived at by scientific investigation alone. There would be no place for a creation; the universe was unchanging, infinite and eternal and would simply always have been there. And man, with all his prejudices, religious beliefs and sense of self-importance, would just be one of the astonishing consequences of a scientifically explained phenomenon. A few scientific thinkers began to profess atheism, belief in the nonexistence of God, as the only intellectually justifiable faith.

The Doppler effect

Ironically, just when some thinkers were beginning to believe that science would bring about the end of religion, physics gave them a reason to pause for thought. The next significant scientific discovery about light would be seen as supporting the creationist convictions of the Church rather more than the arguments of atheists. Not that anyone could have predicted it at the time. Unlike von Fraunhofer, Christian Doppler did recognize the significance for astronomy of what he discovered in Vienna in 1842; but it was to be another seventy years or so before it revolutionized cosmology and appeared to give credence to the Church's views on creation.

Doppler discovered a principle which applies equally to light and sound; it is perhaps easier to grasp the basic idea by considering sound first. What is now known as the Doppler effect is frequently represented by the sound of a train rushing into a station and out again. Anyone standing on the platform will hear the sound of the train change as it approaches, passes, and then goes away again. Obviously the sound gets louder as the train gets nearer and quieter as it gets farther away, but there is a change in pitch as well. The approaching sound is higher-pitched, and the departing sound distinctly lower-pitched. Yet, for somebody sitting

Christian Doppler's effect proved an invaluable tool for analyzing starlight, enabling us to calculate the speed and direction of movement of distant stars and galaxies.

on the train, there is no change. What is the explanation for the change of pitch experienced by the observer on the platform?

Doppler realized that the sound changes because of the changing time it takes for the sound to reach the observer on the platform. It becomes clearer if the train's journey is broken down into a series of moments. Suppose at moment "A" the train is 100 meters (110 yards) away from the observer and approaching fast. The sound of the train will have to travel that 100 meters before it is heard. This takes only a fraction of a second, but it is a finite time. Call it 300 milliseconds. By then the train is, say, only 90 meters (100 yards) away from the observer, at moment "B." The sound of the train will again have to travel to the observer before it gets heard; but now it only has 100 yards to travel. Since sound always travels at the same speed, this will take only 270 of our milliseconds; 10 percent less time to travel 10 percent less distance; 30 milliseconds faster than from moment "A." And, in the same way, the sounds from subsequent moments, "C," "D," and so on, will progressively take less time to reach the observer. This means that the sound of the approaching train reaches the observer all "squashed up."

Once the train has passed the observer, the opposite happens. Imagine now that the train is 100 meters away from the observer at moment "X." The sound from this moment "X" has to travel 100 meters to reach the observer, which it takes 300 milliseconds to do. By this time the train is now, say, 10 meters (11 yards) farther away from the observer, at moment "Y." This means that the sound from moment "Y" has to travel 110 meters (120 yards) to reach the observer, taking 330 milliseconds: 30 milliseconds slower than the sound from moment "X." So now the sound is arriving all "stretched out," while the train gets farther and farther away.

This analogy of trains rushing through stations is so far removed from cosmology that it might seem hard to connect the Doppler effect with light, stars, and the creation of the universe. In fact Doppler first associated the effect now named after him with light, not sound; but it is much harder to understand the principle with light, because we simply cannot experience it in everyday life. The Doppler effect is, however, the key to understanding the new picture of the universe which emerged at the beginning of the twentieth century. And, once you are comfortable with the idea of sound "squashing up" as you observe something moving toward you, and "stretch-

---

The most famous experiment to confirm the Doppler effect was conducted by Christopher Buys-Ballot in Holland. He put a group of musicians on a train and took up his position on a station platform. He asked the train driver to rush past him as fast as he could while the musicians played and held a constant note. Buys-Ballot was able to detect the Doppler shift—a change in pitch as the train passed him. *Inset* This shows how the same effect can be seen in light from a galaxy. If the galaxy stays at a constant distance from the Earth, Fraunhofer lines in the spectrum produced from its light waves will appear in the "standard" position (*top*). If the galaxy is moving away from us, the waves will seem stretched and the Fraunhofer lines will be red shifted (*middle*). And if the galaxy is moving toward us, the waves will seem squashed up, and the lines will be blue shifted (*bottom*).

ing out" as you observe something moving away from you, you are almost there. Another way to visualize it is to picture the peaks and troughs of a sound wave being squashed closer together to produce a higher-frequency, higher-pitched sound; or stretched further apart to give a lower frequency and pitch.

Light, of course, also travels in waves—at very much higher speeds than sound. Light and sound are also both forms of energy. We are all familiar with the way lightbulbs and hot plates give off light and heat when we use their energy to see by or to cook. The direct applications of sound energy are perhaps less commonly experienced, though words like "sonar" and "depth sounding" reveal some of them.

We tend to think of light as being exclusively the light we can see with our eyes; but this visible light is only a tiny part of a whole spectrum of different wavelengths of energy. When Newton created a rainbow of refracted light through his prism, he was separating visible light into its different wavelengths or frequencies. Red, at one end of this rainbow spectrum, is a lower frequency than blue at the other end. Above the blue or violet end of the visible light spectrum there are even higher frequencies or shorter wavelengths, starting with ultraviolet light, and then X rays at shorter wavelengths still. Below the red end of the spectrum of visible light there are longer wavelengths or lower frequencies, starting with infrared light which can be detected on a photographic plate; and then microwaves; and then radio waves, which are used to carry radio and television signals. Visible light is simply a small band within a wide range of waves which are all electromagnetic manifestations of energy, detectable and exploitable in a whole variety of ways, from hospital X rays to microwave ovens.

This range of electromagnetic waves is also very useful for astronomers who want to observe the early universe. We are all familiar with detecting heat and light; many people are aware that X rays and infrared light can be detected on a photographic plate. It is perhaps less well known that scientists can detect other parts of the electromagnetic spectrum, even when there is only a minute amount of energy being radiated. The waves can be detected long after everything has cooled down to almost nothing; in just the same way as you can feel the last shimmer of heat from the ashes of a dead fire. This means that astronomers can detect evidence of events that were extremely hot once, but have cooled to almost nothing millions of years later. Sometimes the source of the radiation can be so far away that, even though the waves travel at the speed of light, they take billions of years to reach us. Through these light waves, we can experience what things must have been like billions of years ago. This applies to visible light too, which is why we speak of images seen through powerful telescopes as being a certain number of light years away. Saying that something is a number of light years away simply expresses the distance the star or galaxy is away from us, measured by the time it takes for the light to reach us.

Because light from distant stars has so far to travel it can also reveal the Doppler effect. Although Doppler realized that the effect he had discovered would apply to light waves as much as sound waves, it was difficult to test his theory with a simple experiment on Earth. As far as visible light was concerned, because light travels so fast, he reasoned that it would be hard to observe the Doppler effect unless the source of the light was at a great distance from the observer, and moving very fast. Otherwise any "squashing up" or "stretching out" of the waves would be so minute as to be imperceptible. He reasoned that stars were about the only sources of light far enough away from us for the Doppler effect to register—if, that is, they were moving away from us or toward us at a fast enough speed. He chose to study a pair of stars which astronomers thought from their observations might be sharing the same orbit. Doppler reasoned that, if they were, at any one time one star would be moving away from us while the other moved towards us. It's a bit like two spots on the opposite edges of a spinning top: when one spot moves through the point in the spinning cycle nearest someone watching, the other spot has to be moving through the point in the cycle furthest away. And, as the first spot moves away from the person

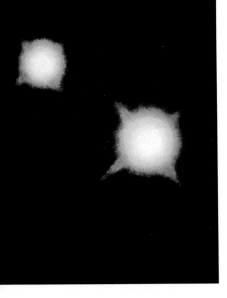

The two stars of Alpha Centauri have now been observed often enough to establish how they move in relation to each other. Analysis of their light over a period of time shows a series of changes in their spectra. The way the red and blue shifts vary gives us a picture of two stars completing orbits around each other once every 80 years.

watching, so the other spot starts to come toward him.

Doppler calculated that, when studying the light from each star, if they were indeed in orbit, he should see a difference in the wavelengths which would tell him which way each star was moving. He carefully refracted the starlight from each one, and saw the characteristic Fraunhofer lines for hydrogen and helium in each of the spectra. But there was one crucial difference. The identical line pattern in each spectrum was in a slightly different position. It was as if one had been shifted toward the blue end of the visible light spectrum, and the other shifted toward the red end.

This was exactly as Doppler had predicted; one light source was being seen at the higher frequency of the blue end of the visible light spectrum, consistent with a "squashed up" light wave; and the other, at the lower frequency of the red end of the spectrum, which you would associate with a "stretched out" wave. The blue shift was, in effect, the light "squashed up" as the one star moved toward him, just like the "squashed up" sound of an approaching train. The red shift was the "stretched out" wave from a star moving away from him, just like the "stretched out" sound from a departing train.

In other words, the direction of the movement of stars could be detected by analyzing the light from them, to see if it was red or blue shifted. And the farther the Fraunhofer lines were shifted toward the red or blue end of the spectrum, the more "stretched out" or "squashed up" the waves had to be; in other words, the faster the star was moving away from us or toward us. Suddenly, by analyzing light, Doppler had a means of determining both the direction and speed of travel of every source of light in the night sky.

While astronomers acknowledged the Doppler shift as a new tool for them to use in exploring the universe, there was no immediate sense that anything truly revelatory had happened. Certainly there was nothing yet to rekindle creationist enthusiasm among churchgoing scientists, or in any way threaten Newton's model of an infinite, eternal universe. What Doppler saw merely confirmed what astronomers had already expected: the stars were moving. Newton had predicted a universe full of moving heavenly bodies; and, ever since early Greek observations, long before Galileo had seen the moons around Jupiter, much of what moved was perceived to be in orbit. And yet the use of the Doppler shift in studying starlight was to have as revolutionary an effect on the understanding of the universe as Galileo's observations or Newton's theory of gravity. What was required was a master craftsman to make full use of the tool; and he did not emerge for another seventy or eighty years.

Mapping the galaxies

When he did, he had a personality large enough for the part. Edwin Hubble, born in 1889, was a young American who thought about a career as a boxer and took a Ph.D. in Law at Oxford before finally deciding to become an astronomer. Some of his contemporaries in the 1920s at the Mount Wilson Observatory in California remember him as a meticulous observer, patiently and precisely putting every piece of observational evidence together, rather like a careful prosecuting lawyer preparing his case. Others recall the instinctive intellectual inspirations of a man who was to see the meaning of everything in a moment of instant insight. Perhaps he was in fact both.

Hubble managed to get the use of the most powerful optical telescope of his time, on top of Mount Wilson. The first thing he wanted to do was to study the distant galaxies, and see if he could learn anything more about them from examining the light they radiated, using the Fraunhofer fingerprints and the Doppler shift to try to pinpoint their movements and chemical makeup. And he had another useful tool with which to establish how far away they were from Earth.

Before the twentieth century, the basic method of calculating how far away stars were from the Earth had remained remarkably little changed for two thousand years. Methods had, of course, become more sophisticated since Eratosthenes and the early Greeks had used sticks and geometry to work out the distance to the sun; but it was really still only possible to refine these calculations to a limited extent. Mathematics had progressed sufficiently to calculate the distances to nearby stars in our galaxy, and no further. But in 1912 an American astronomer, Henrietta Leavitt, had spotted one kind of star which revolutionized the way astronomers measured distances. Stars do literally twinkle; many of them vary the intensity of the light they give out from one time period to the next. The reasons are complex and less important than Leavitt's realization that the variability of the intensity of the light from one kind of star was uncannily predictable. They were dubbed Cepheid stars, after the position in the galaxy where one was first observed.

There are comparatively few Cepheid stars in our galaxy; hundreds compared to billions of other kinds of stars. But, once one is spotted, its

---

Edwin Hubble discovered that other galaxies were an astonishing distance from our own. This is the Whirlpool Galaxy, some 13 million light years away. In other words, by the time light from the galaxy has reached us, we are seeing how the galaxy looked 13 million years ago.

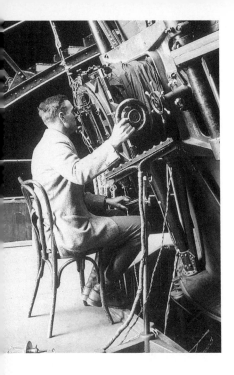

Edwin Hubble (1889–1953) at work observing with the 100-inch Hooker Telescope (*right*) on Mount Wilson, California. The "100-inch" part of this name refers only to the size of the mirror in the telescope; you can get an idea of its overall dimensions from the chair on which Hubble sat. You can just see that it is still in place on the viewing platform, at the extreme right of the main telescope picture.

variable light pattern is so marked—and so different from other stars—that it is unmistakably a Cepheid. It is a bit like spotting a yellow tulip in a field of red ones. And, because it is so predictable, it will have an exact level of overall brightness in relation to how often it switches from its most intense to its least intense light. It was this which gave Hubble a tool with which to measure the distance from the Earth to the farthest corners of space. You can work out with traditional mathematics how far away a nearby Cepheid is, and also measure its luminosity or brightness. This becomes a yardstick with which to make comparisons. Say you are trying to discover the distance between Earth and a far-off galaxy. First you have to detect a Cepheid star in this galaxy. Then you can measure the luminosity of this more distant Cepheid and compare it to the light from the nearby Cepheid star, at a known distance from the Earth. The proportionate difference in brightness will correspond to the proportionate difference in distance from the Earth.

This meant that, as long as powerful telescopes could pick up the unmistakable variable light pattern of a Cepheid star in a distant galaxy, its distance away from Earth could be calculated—Hubble now set about the daunting task of mapping the universe galaxy by galaxy. He realized that he could not only check how far away each was from Earth, by comparing a Cepheid star in it to one in our own galaxy; he could also learn what each galaxy was made of, using the Fraunhofer lines in the refracted light to identify the elements in its stars. And he could also get an idea of the direction in which the galaxy was moving, and at what speed, from the amount that the fingerprint was Doppler-shifted in the refracted light. The farther the light was shifted, the more "squashed up" or "stretched out" its light would be, and the faster it had to be moving. He and his team patiently iso-

As the roof of the observatory opens, the telescope is lifted into position and the viewing platform travels up and along the curved track to be alongside the top of the telescope, where the eyepiece is. Light from the star being observed reaches the mirror at the bottom of the tube and is reflected back up to the eyepiece at the top.

lated the light from each distant galaxy, refracted it and analyzed it, and so built up as complete a picture of it as they could.

Not surprisingly, they found that hydrogen and helium were the most abundant elements making up all the galaxies. This merely confirmed what William Huggins had seen. What was more surprising was the fact that all the light they were analyzing was red shifted. In other words, every galaxy seemed to be moving away from us. And, from the Cepheid stars they identified, they found that the galaxies were even farther away from us than anyone had ever imagined. Some were billions of light years away; so what Hubble and his team were seeing was light that had traveled for so long that it revealed what these galaxies were like, and so how far they had evolved, some eight thousand million years ago. And this light was red shifted farther than any other galaxies. The oldest and farthest galaxies, in other words, were traveling away from us at a phenomenal speed; faster than the galaxies nearer to us.

These spiral galaxies were selected from among his observations by Edwin Hubble as examples of the genre in 1925. Hubble classified galaxies as elliptical, irregular and spiral; his classification still remains broadly accepted today.

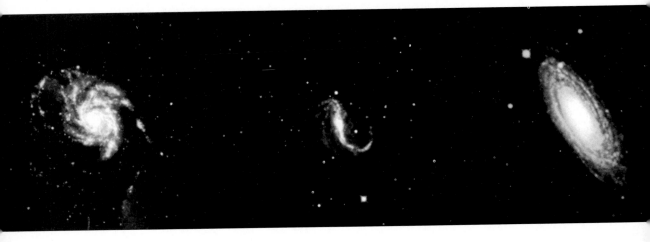

The expanding universe

What Hubble had discovered was so unexpected that its full significance was not immediately grasped. There are not many dynamic systems in which everything is moving away from you, wherever you look. One good example is an inflating balloon. Suppose you mark a spot anywhere on its surface, and then mark any number of arbitrary spots all around it. As you blow the balloon up and its skin stretches, so all the spots on its surface will get farther away from the first spot you marked. In other words, like an inflating balloon, the universe had to be in some way expanding.

It is extraordinary to think that this dynamic property of the universe could be discovered entirely through the analysis of starlight. It was such a startling new finding that many physicists' first reaction was to question whether Hubble's interpretation was correct. There had to be another explanation, they felt. But Hubble was able to show a constant relationship between the speed of a galaxy's movement, as revealed by its red shifted light, and its distance from Earth as measured by the brightness of the light from its Cepheid stars. His observations were uncannily consistent; everything at a given distance from the Earth was moving at the same speed. And the greater that distance was, the more the speed increased. In fact Hubble was able to write a mathematical equation expressing this precise relationship—Hubble's Law—which turned out to fit perfectly every time new data was gathered from a new galaxy.

An expanding universe was a difficult concept for the majority of atheistic scientists who had become firmly wedded to the idea of an unchanging infinite and eternal universe. Anything that was expanding could hardly be unchanging. So there was a huge temptation to play down or dismiss Hubble's perception. But the new discovery did excite at least one group of scientists; and, in particular, one visionary astronomer. He was a priest in the Vatican whose views offered perhaps the most dramatic explanation for Hubble's picture of an expanding universe.

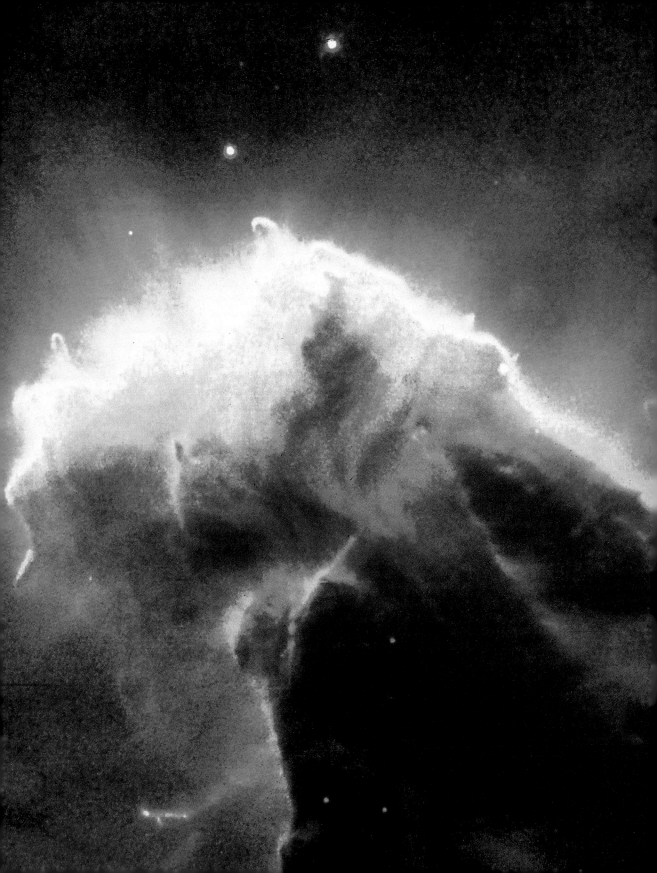

~~~~~~~~~

In the Beginning . . .

IN 1927, GEORGES LEMAÎTRE, A BELGIAN JESUIT PRIEST AND THE leading theoretical cosmologist working at the Vatican Observatory, had been musing over some of Albert Einstein's ideas and mathematical equations. Lemaître himself vehemently argued that he simply wanted to make a model of the universe which would be consistent with Einstein's theories; but others were convinced that he was trying to come up with a scientific explanation of the universe which would allow for a moment of creation: something which Newton's eternal and infinite model seemed to rule out. It was

---

Spectacular clouds of dust like the Eagle Nebula, billions of miles across and rich in hydrogen molecules, are now seen as the typical birthplace of stars. Just one small piece in the incredible jigsaw which had to be pieced together after Hubble saw that the universe was expanding.

important for the Catholic Church to find a way of making the creationist ideas in the Bible consistent with all the scientific discoveries about the universe; and its scientists in the Vatican Observatory were finding this an uphill task. Lemaître apparently wanted to find some new evidence to suggest that the universe was finite and therefore must have had a clear beginning.

It's all relative

He was working through Einstein's theorems because Einstein was rapidly becoming known as the visionary scientist of his time. Einstein had worked as a patent officer in Berne, Switzerland, to earn a living and pay for his academic work while he wrote up his ideas about the laws of physics. His first major work was published in 1905, the first of two Theories of Relativity. It is called Special Relativity; and the later theory, published in 1915, is called General Relativity. Both deal with the way an observer and the event he or she observes are related; Special Relativity essentially spells out what happens when there is a constant movement linking the event and the observer, and General Relativity brings in gravity. It also suggests what happens as the speed of any movement increases or decreases. They are both still very difficult theories to understand fully, but they are nevertheless widely acknowledged as the ideas which placed Einstein on the scientific world stage. Einstein did not set out specifically to explain the universe, but his theories inevitably interested cosmologists, because he was in effect rewriting the laws of physics which had been left unchallenged since the time of Newton.

Einstein argued that the laws of physics must be the same, from whatever position they happened to be observed. This idea stemmed from the insight that the same event can appear different to two different observers, depending on their relative positions. Several day-to-day examples have been suggested to help illustrate the point. One that most of us have experienced at some time or another is when two trains stop alongside each other in a railway station. You can be sitting on one train, looking out of the window at the other train, when it seems to move off. For a second or two you are not sure whether it has in fact started to move, or whether it is your own train which is moving off. All you know is that one train must be moving relative to the other; hence Relativity.

Albert Einstein (1879–1955) never made it his direct ambition to solve the cosmological riddles of the universe, but his revolutionary theories radically affected cosmology as much as any other branch of physics.

Génia Reinberg

Now imagine a situation where one observer is on board a train and another is on a railway station platform as the train rushes by. A cup on a table in front of the man on the train will appear to stay 60 centimeters (2 feet) in front of him. So, from his point of view, it will not be moving. However, to the man on the platform, the cup will be seen (through the window of the train) to rush past at great speed as the train hurtles through the station.

Einstein's great insight was that the laws of physics had to be rewritten in such a way that the laws of motion would be recognized as being consistent. They would have to account for related concepts such as acceleration and momentum, which were involved in these apparently different views of the cup. And this meant understanding the nature of time and space, and how they affected things. After all, what causes the two different views of the cup are the different positions of the two observers relative to the cup in time and space. One is traveling through time and space alongside the cup, so that its relative position is always 60 centimeters in front of him; it stays in his field of vision as long as they are both traveling through time and space in an identical fashion. The other observer is, by comparison, stationary in time and space relative to the moving cup, so that it comes into and moves out of his field of vision in a very short time.

Sinking into space-time

Einstein developed mathematical equations to describe these kinds of relationships. Taken together, they defined the nature of time and space; and they had momentous consequences for cosmologists. To begin with, it emerged that time and space were mathematically one and the same thing. And, as a consequence, Newton's explanation of gravity had to be totally revised, accurate as it seemed to be. Einstein argued that two objects do not directly attract each other as Newton had thought; rather, each of the two objects affects time and space, and any gravitational effects are a consequence of this. If this concept is difficult to grasp, imagine a heavy object (such as a cannonball), representing the sun, being placed in the middle of a taut rubber sheet which represents space-time. The weight of the ball will

---

Einstein's theory of gravity is often demonstrated by imagining heavy objects, representing stars or whole galaxies, in place on a rubber sheet marked with grid lines to represent space and time. The more massive the object, the greater the dent it will make in space-time and thus the harder it is for anything passing nearby to avoid being pulled into the object.

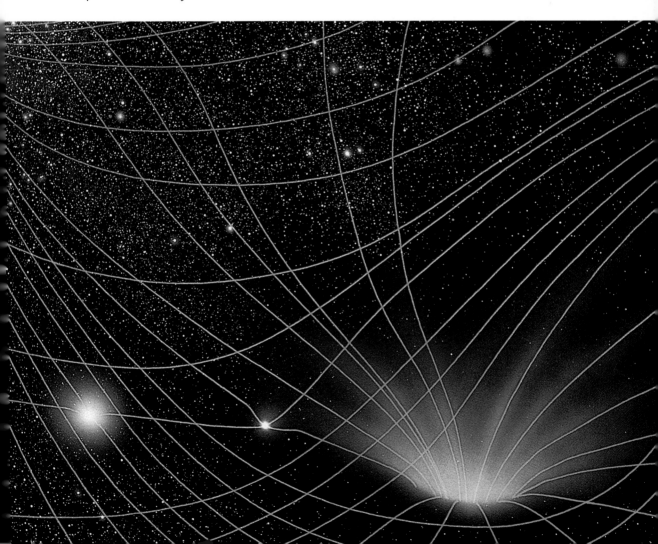

Abbé Georges Lemaître (1894–1996), a Catholic priest and Belgium's most famous astronomer, loved good food and wine, and was a popular figure at the Pontifical Academy of Sciences. But scientists elsewhere were slow to pay attention to his "primeval atom" theory.

make it sink into the rubber sheet, creating a cone-shaped dent all around it—rather reminiscent of the surface of a vortex of swirling water rushing down a plug hole.

Einstein argued that whenever something heavy bent space-time like this, it would naturally affect the path of anything lighter traveling nearby. So a small ball representing the Earth or one of the other planets could be rolled across the stretched rubber sheet representing space-time, toward the dent around the cannonball sun. If it was traveling too slowly, it would fall directly into the dent and quickly reach the surface of the sun (just like Newton's apple falling to the surface of the Earth). If it was traveling too fast, it would have its path deflected towards the cannonball sun, but would only dip into the dent then climb out the other side, before continuing on its journey. But at just the right speed, the small planet ball would be traveling fast enough not to fall right into the dent, but too slowly to escape it completely. With nothing else to stop it or slow it down, it would find its level on the "side" of the dent, thus orbiting the sun ball—rather like a motorcycle stunt rider going around and around the "wall of death."

The mathematical formulas which described gravity in this way gave very similar results to Newton's rather simpler equation; but they also exactly fitted the orbit of Mercury around the sun, which you may remember Newton's formula did not predict perfectly. This was impressive evidence that Einstein's theory was correct, or at least an improvement on Newton's explanation of gravity. And an experiment we will come to later, involving the eclipse of the sun and starlight, was also to confirm the accuracy of Einstein's predictions. It was natural for physicists to begin to think: if it fits in with Einstein's theories, it is probably going to be true.

It was while studying these equations of Einstein's that Lemaître discovered something which really excited him. One of the consequences of Einstein's mathematics was that the universe was not static; it was dynamic. It is simple enough to see why. If time and space are "dented" by anything with mass, then, as one heavenly body passes another, it will be drawn closer to it. If the universe is static, then all objects will eventually be drawn to each other; all mass will congregate together at the bottom of the largest dent in space and time. This was the same problem which had worried Newton when he came up with his theory of gravity; how could all the matter in the universe still be widely spread out after billions of years? Why hadn't it been pulled together by gravity into one conglomerate lump? But, whereas Newton's idea had confined itself to the attraction of objects, Einstein's theory involved the mathematics of how space and time change when an object with mass affects them. Thus Newton's system had no way for the coming together of all objects to be avoided but Einstein's mathematics did. Einstein needed space and time to be able to change in the presence of mass. So space and time had to be dynamic, rather than static. Consequently, space-time, and so the universe, could not remain still; and if it had to change it could only really get bigger or smaller. Hence it ought to be gently expanding or contracting.

Einstein's red herring

Einstein had spotted this himself, and was unhappy about it. As a firm believer in the Newtonian idea of an infinite unchanging universe, he convinced himself that there was a law of physics which would prevent any overall expansion or contraction. There had to be something which would allow local variations, as space-time was affected by mass, but not affect the overall status of the universe. Einstein therefore added an extra factor to his equation, the "cosmological constant": a kind of weakly repulsive force to cancel the inward pull of gravity and thus prevent an overall dynamic change.

But Lemaître could see no reason to introduce this "cosmological constant" out of thin air. Supposing you just stayed with the mathematical model of a gently expanding universe. This would mean that the expansionary force would counter the gravitational force, and so all the matter in the universe would stay separated. Not only that; if the expansionary force slightly exceeded the gravitational force, then the universe would continue to expand and become bigger tomorrow than it is today. That would also mean that it had to be smaller yesterday than it is today, in order to expand to today's size. The universe would therefore have to be progressively smaller the further back in time you went. And that meant that at some point, very long ago, the universe would have been at its smallest possible size.

Lemaître suggested that this would be the starting point of the universe, the moment of creation his Church had been looking for. He thought he had found the perfect model: a universe which God had created as a "primeval atom," which continued to grow and expand like an oak tree from an acorn; and a universe which at the same time faithfully followed all the mathematics of Einstein, the scientific guru of his day. It was also a universe which coincidentally solved the problem Einstein had had with the expansion his unaltered equations had predicted.

Unfortunately for Lemaître, Einstein remained emphatically unimpressed. He suggested that Lemaître had a poor grasp of the physics involved, and implied that it was "obvious" that the universe must be infinite, eternal and unchanging. It was ridiculous to suggest a moment of creation from a primeval atom. And if Einstein was so sure it was wrong, who else in the scientific community was going to believe in Lemaître's rather extraordinary idea?

The Catholic Church, however, was naturally delighted with the idea, and Lemaître was encouraged to persist with it. Within two years, Lemaître heard the news he had scarcely dared hope for. There was other scientific evidence to suggest that the universe was expanding. Hubble had observed that the light from galaxies was red shifted, and, according to the Doppler effect, this had to mean the universe was expanding.

Now it was only a matter of time. Einstein was interested in Hubble's work anyway and resolved to visit him at the Mount Wilson Observatory. Lemaître arranged to give a lecture at the California Institute of Technology at the same time, and managed to corner Einstein and Hubble together. He argued his "primeval atom" theory carefully, step by step, suggesting that the whole universe had been created "on a day which had no yesterday." Painstakingly he worked through all the mathematics. When he had finished he could not believe his ears. Einstein stood up and

Albert Einstein (standing on the left of the group) visited the Mount Wilson Observatory where Edwin Hubble had made his historic observations; and then met with Hubble and Lemaître to discuss their ideas back down at sea level. It was at this meeting that Einstein realized the mistake he had made with his "cosmological constant."

In the Beginning . . .

announced that what he had just heard was "the most beautiful and satisfying interpretation I have listened to" and went on to confess that creating the "cosmological constant" was "the biggest blunder" of his life.

This was a great triumph for the Catholic Church, which now had a model of the universe that included a moment of creation. It was a model which fitted the biblical picture as well as having the support of the greatest scientist of the age. Moreover, by working carefully with Hubble's data, it was even possible to suggest the timescale involved. By calculating the speed at which galaxies were traveling (from the extent to which their light was red shifted) and by knowing how far they were from the Earth and each other at different points in time (using the data from the Cepheid stars in each galaxy), it became possible to project back to a time when all the galaxies had been crushed together at the same point. This would have been Lemaître's moment of creation; and it turned out to have been about fifteen billion years ago. This radically different picture of a dynamic universe growing from a specific starting point clearly suggested that the idea of an infinite unchanging universe might be completely mistaken.

But several eminent scientists—especially the committed atheists—were still not convinced. The idea of the universe somehow growing from something smaller than an atom seemed too preposterous for them to take it seriously. One influential group at Cambridge set about finding an alternative explanation. Why not suppose, they argued, that we are just not seeing the whole picture? The universe we can see may be expanding, but other parts, beyond our field of vision, might be moving in the opposite direction and contracting. Overall, the universe could be in a steady state, embracing pockets of expansion and contraction, bubbling and boiling forever like a giant cauldron of water. After all, they reasoned, Einstein's mathematics would allow for both expansion and contraction.

The life and death of stars

The "steady state" theorists included one of the best-known scientists of the time, Fred Hoyle, a physicist and a confirmed atheist. One of the ideas embraced by steady state theory was that, in our expanding corner of the universe, as galaxies grew farther and farther apart, new stars would evolve to help fill the gaps left by this expansion. Hoyle's great achievement, along with others, was to explain the life cycle of a star which this idea implied.

At the beginning of the twentieth century, scientists were making great strides in understanding the nature of all matter: the various chemical ele-

The steady state theorists at Cambridge University included Fred Hoyle (*left*) and Hermann Bondi (*right*). They were part of a small but influential group who for a while managed to cast serious doubts on the idea of the universe growing from a small beginning.

ments which make up everything in the universe, and the subatomic particles which in turn make up these chemicals. When Hoyle and his colleagues were working out the nature of stars, it was therefore well understood that the primary chemical elements in the universe could only have been forged in places where there was immense heat and pressure—far greater than anything which ever existed on Earth. This pointed to the stars as a possible source of these primary chemical elements. If steady state theory demanded

An artist's impression of some of the consequences of a star's final collapse. In the foreground is a brown dwarf star (*top right*) and two white dwarf stars can be seen still shining in the distance (*upper center and center left*) beyond all kinds of dark matter (see chapter 8). The view is from somewhere just above the edge of a galaxy (*bottom left*). And grid lines have been drawn (*bottom right*) to suggest how a black hole (see chapter 11) would bend time and space.

the birth of new stars to fill the gaps in an expanding part of the universe, they reasoned, why should stars not have a dynamic life cycle, with moments of birth and death, giving them a lifetime in which they produced all the chemical elements? The picture put together by Hoyle and others was that stars formed when hydrogen atoms in space were pulled together by gravity into bigger and bigger spheres. The inward pressure of gravity as the spheres grew bigger (a bit like a rolling snowball gathering more and more snow) would steadily increase. Eventually this pressure would become so great that some of the hydrogen atoms would get pressed together so tightly that they would fuse together to become helium—the next heaviest atoms.

Like any nuclear reaction, this process of atomic fusion involves the release of huge amounts of energy. (The most dramatic example of such an energy release on Earth is when an atomic bomb goes off; but this is a minor explosion compared with the nuclear reactions going on in the stars—something like a balloon bursting compared to the entire Earth exploding.) As the hydrogen is fused into helium atoms, the energy released does two things. Firstly, a lot of it provides an outward explosive pressure to counterbalance the inward pressure of gravity in the emerging star. This allows the star to remain stable for billions of years. Despite all the fusion reactions taking place inside it, the star neither explodes like a bomb, nor collapses under the influence of gravity; a balance is maintained. Secondly, some of the energy produced is not involved in this balancing act, and escapes the star, radiating outward as heat and light. Hoyle's theory for the creation of the elements also neatly solved an ancient mystery; it explained why stars shone.

Eventually, however, Hoyle and the others argued, there must come a time when the hydrogen in the star is almost all used up, and the predominant chemical is helium. With less hydrogen to fuel the fusion reactions, there will be less outward pressure, and the balance with the inward pressure of gravity will be disturbed. As the gravitational pressures build up, the inward gravitational pressures will increase; and eventually they will become powerful enough to press all the newly created helium atoms tighter and tighter together. So the helium atoms will start fusing together, to form atoms of the next heaviest chemical element. And this process will go on, as first one element, then the next, becomes the dominant chemical in the star, and in turn fuels the fusion reactions, until the gravitational pressures build up and press the atoms tightly enough together to form the next heaviest element.

Because the strength of the force of gravity depends on the overall mass of the star (and so, in effect, its size and weight), it is possible to calculate that there will be a difference in the ultimate fate of a small star and a large star. When all the elements up to iron have been produced by the progressive fusion reactions, it needs a huge increase in temperature and pressure for fusion to create the next heaviest elements. And small stars cannot generate these temperatures and pressures. The inward gravitational pressures are simply not powerful enough. The fusion process beyond iron cannot get started, and so the star starts to die. Not all the hydrogen in the star would have been completely used up before the fusion of helium got started; nor would all the helium before the next stage in the chain. So the star will contain some of all the lighter elements up to iron before it goes into its death throes. All these other elements remaining from the earlier fusion reactions are shed into space as the star cools down, leaving only a hot iron core which still shines for a while as what is known as a white dwarf star, before eventually cooling into a brown dwarf: a cold iron sphere which no longer emits light. This final relic of a star will remain forever in space, unless gravity or some other cosmic force pulls it into collision with other bodies in the universe.

The colorful blast wave of a supernova seen by the Hubble Space Telescope. The intense heat from the exploding star can light up the cosmos for billions of miles, as well as creating all the elements heavier than iron.

But with bigger stars, the gravitational pressures are so strong that, even though they cannot start the fusion reactions needed to turn iron into heavier elements, they can crush the iron core of the star until it implodes. This releases so much energy, as the star dies in a dramatic explosion in space, that, for a short time, high enough temperatures and pressures exist for all the heavier elements to be created. And astronomers' telescopes have seen the dramatic supernova, as it is called, which results: a great cloud of hot shining dust blown out into space, spreading all the elements far and wide. (Incidentally, we now know that the same supernova explosions can give rise to very dense neutron stars or pulsars—a kind of exotic new star emitting radio signals at regular intervals. This is confirmation of the extraordinary levels of nuclear activity forged in the heat of the supernova, which are responsible for the creation of the heaviest elements.)

In fact, all that Hoyle and others predicted has been confirmed by observation. Using the lines in the spectrum of refracted light which von Fraunhofer had discovered, scientists were able to analyze the light from supernovas and detect the chemical elements which were present. As Hoyle and the other steady state theorists had predicted, all the heavier elements were there. Besides unlocking the secrets of supernovas, telescopes fitted with prisms to refract the light they were observing soon recorded stars with all the light fingerprints of the elements up to iron. And then they sometimes saw their light levels diminish before they finally, in effect, disappeared. Furthermore, as the understanding of the mathematics of nuclear physics gets more and more sophisticated, so the inevitability of the fusion reactions is more and more clearly predicted. So there is no real doubt that Hoyle's explanation of the life cycle of a star is correct.

At first sight this incredible work gave support to steady state theory, since it showed how stars could not only come into being but could also provide all the elements in the universe. It seemed to paint a convincing picture, at least for the atheists, who wanted to find a way for everything in the universe to come into existence without the need for a moment of creation, or Lemaître's primeval atom. For the time being, at least, it was possible to conceive of an infinite, eternal universe with an internal dynamic system to explain the red shifts which Hubble had seen. But, brilliant as the explanation of matter forming from stars was, it had within it the seeds of its own destruction. Steady state theory was not destined to offer significant opposition to the idea of the Big Bang for very long.

chapter
five

~~~~~~~~~~~~

Relics, Singularities and Ripples

THERE IS NO DOUBT THAT THE EXPLANATION OF THE LIFE CYCLE OF STARS was a colossal intellectual achievement. It meant that the "steady state" ideas of Fred Hoyle and his colleagues won closer attention, and for a while offered a serious alternative to Big Bang theory. It certainly meant that Hoyle was well enough respected as a scientist for Stephen Hawking to hope, in the early 1960s, that he would supervise his postgraduate doctorate at

---

Countless stars jostle around the constellation of Musca in a densely populated area of our galaxy, the Milky Way. Just one part of one of billions of galaxies, each a unique mixture of energy, movement and beauty. It is not easy to see how so much variety could come from a single explosive beginning to everything.

Cambridge. As it happened, Stephen had his Ph.D. supervised by another steady state supporter, Dennis Sciama, who eventually came to support the thrust of Lemaître's ideas about the origins of the universe. Hoyle, to this day, apparently prefers the steady state theory.

The Big Bang comes of age

A number of things gradually persuaded people like Dennis Sciama to change their minds. Perhaps the first and most telling blow to steady state theory came from its supporters' own explanation of how stars cooked up the elements. It was all very well to suggest that all the elements could be derived from hydrogen in the life cycle of the stars but that raised an obvious problem: where had the hydrogen come from to make the stars in the first place? It would need, according to subatomic theory, a colossal high-temperature explosion to create hydrogen from subatomic particles. One possibility was that hydrogen would somehow have been forged in the enormous heat which exploded from Lemaître's primeval atom (the very concept which the steady state theorists wanted to dismiss). In a radio broadcast at the end of the 1940s, Hoyle had brushed this problem aside and asserted that it had to be explained in some other way. "If the universe began with a hot Big Bang," he mocked, "then such an explosion would have left a relic. Find me a fossil of this Big Bang."

The name he coined in trying to denounce the theory has, ironically, stuck ever since. And his challenge to its supporters to find a relic of the Big Bang has led to the discovery of a great deal of evidence in support of it. Far from killing it off, Hoyle managed in the long run to promote the Big Bang theory. In fact in 1948, almost as soon as Hoyle and his Cambridge colleagues, Hermann Bondi and Thomas Gold, first formally announced their arguments in favor of a steady state universe, the case against them was being put together by another group of physicists. For poor Fred Hoyle, 1948 turned out to be memorable for all the wrong reasons.

Cosmic interference

For a start, in 1948 George Gamow and his student Ralph Alpher worked out that, if there was a Big Bang, Hoyle's fossil should exist. They calculated

Ralph Alpher (*left*), Hans Bethe (*center*) and George Gamow (*right*) published a joint paper in 1948, to which Bethe had made no contribution. Gamow added Bethe's name for a joke; the paper became known as the Alpha, Beta, Gamma paper. But Bethe was more than just a convenient name; he won the Nobel Prize for Physics in 1967 for his work on the energy in stars.

David Wilkinson at Princeton University with the horn antenna he helped build.

that the Big Bang would have generated an incredible amount of heat in order to create the hydrogen needed to form the first stars. In fact it would have created some helium as well as the hydrogen. Their prediction was 80 percent hydrogen and 20 percent helium, which is exactly the proportion that was detected from the Fraunhofer fingerprints seen in the light from the earliest galaxies. They argued that the afterglow of the intense heat which created these elements would not have completely disappeared, even after billions of years.

Since the universe had expanded in all directions from the Big Bang, they reasoned, then you would always be able to detect this low-level background radiation in whatever direction you looked. There would not exactly be much warmth left; their calculations put it at only a few degrees above absolute zero, the lowest possible temperature, or $-273°$ centigrade! But it should be detectable; and it would be the only radiation that would be spread around evenly wherever you pointed a detector sensitive enough to pick it up. Any other radiation generated after the Big Bang would have a specific starting point within the universe; so it would be moving outward from that point only. The radiation from any explosion which started the whole universe, however, could not be traced back to any such single point. It would have to be spread everywhere by the general dynamic expansion of such a universe.

Gamow had quite a sense of humor. He may not even have told him, but he added the name of a third physicist, Hans Bethe, to the paper solely so that it could be authored by "Alpher, Bethe and Gamow," a play on the

The gigantic horn antenna at Bell Laboratories, just down the road, which made the discovery by accident. Bob Wilson and Arno Penzias (*right*) got the Nobel Prize in 1978 for their unintentional historic finding back in 1964.

first three letters of the Greek alphabet. Gamow knew Bethe well enough to be sure that he would not take the joke amiss. But he could not have anticipated the bigger joke that was to follow. It was a classic piece of scientific irony emerging spontaneously from reality, a turn of events which no practical joker could have contrived any better.

In the 1960s, a team at Princeton University was busily engaged in trying to find the background radiation Gamow's paper had predicted: the "relic" which Hoyle had felt sure could not be found. They were working slowly and methodically toward setting up their experiment. They had to have instruments calibrated carefully enough to distinguish the precise low-level radiation they were looking for from all the other radiation in the universe. To do this, they had to add what was called "a cold source"—which

Relics, Singularities and Ripples

had a known temperature against which to compare the temperature of the radiation they were detecting.

Robert Dicke and his team were confident that they had almost completed the ideal test set-up, and were apparently discussing, over a sandwich lunch, what refinement to try next, when the telephone rang. Dicke answered it. It was a call from two Bell Laboratories researchers down the road from Princeton, who were trying to clean up the reception on one of their receivers, partly in preparation for when satellite transmission of broadcasting signals began. They seemed to have run into a bit of a problem. After cleaning everything scrupulously—even scraping some pigeon droppings from where they had removed two nesting birds from inside the giant horn on their receiver—they still could not get rid of a persistent level of interference. They too had added a cold source to their detector, in order to establish the temperature of the ubiquitous interference. They had been told that Dicke and his team were reputedly some of the most knowledgeable people about radiation in space at that time; so they were calling to see if they had any idea what could be going on.

Arno Penzias and Robert Wilson, the two research engineers from Bell Laboratories, had done a thorough job of noting down the characteristics of the troublesome interference. The talk among Dicke's colleagues died down as they began to piece together what was being said, from the side of the telephone conversation they could hear. Yes, they were checking it against a cold source. Yes, it was seemingly always at 3° kelvin—just 3 degrees above absolute zero. Yes, it was in every direction they looked. Dicke's face dropped, and so did the rest of the team's. As he put the phone down, Dicke declared, "Well, boys, we've been scooped." They had been beaten to the discovery they had hoped might win them a Nobel Prize. Two other highly capable academics, employed as researchers by the telephone company, had accidentally stumbled across what the academic team next door was looking for: Hoyle's fossil of the Big Bang. In the end it was Penzias and Wilson who were to win the Nobel Prize.

Falling into a hole

Whoever had found the radiation, it was pretty good scientific evidence in favor of the Big Bang. And, since it happened while he was in the middle of his doctorate, it was perhaps one of the things which inspired Stephen Hawking to focus his Ph.D. thesis on Big Bang theory and Einstein's Theory of Relativity. You will remember that when Einstein had conceded that his "cosmological constant" was a big blunder, in the light of his meeting with Hubble and Lemaître, it left his equations predicting a gently expanding universe. And this had coincidentally provided an answer to one of Newton's problems with his model of the infinite unchanging universe: how movements in the universe could be governed by gravity, which they undoubtedly seemed to be, and yet why the universe as a whole did not eventually get pulled together into one conglomerate lump. Now there was an answer; the expansionary force detected by Hubble's observations and predicted by Einstein's mathematics would counteract the gravitational pull in the opposite direction.

This convinced several theorists that they should follow through the predictions of Einstein's mathematics of relativity, to see what other implications for the universe they might hold. Because Stephen Hawking wanted to pursue something in this area as the central topic of his thesis, Dennis Sciama

After graduating at Oxford, Stephen Hawking went to Cambridge to start postgraduate work in 1963.

Roger Penrose (standing on the running board) was persuaded by Dennis Sciama (next to him) to apply his exceptional mathematical abilities to cosmology. At the same time, Sciama was supervising Hawking's work; and so he became the catalyst for his collaboration with Penrose.

went to hear what another mathematician, Roger Penrose at Oxford University, was making of Einstein's theorems. He was working on another prediction of Einstein's mathematics, which was that gravity would make huge amounts of matter collapse inward into an increasingly smaller and denser point, which he called a "point of singularity."

Scientists were able to accept that, under certain conditions, this made theoretical sense. But when those conditions did not apply, if the collapse happened, it would end up making a nonsense of the very laws of physics which had predicted the collapse in the first place. So, in order to preserve the laws of physics, scientists hoped that there would be a mathematical way to predict an alternative to the collapse. Instead of which, Roger Penrose came up with a mathematical proof that the collapse was inevitable! To most of the general public this all sounded too ludicrous to be true. It was difficult enough for the scientists to envisage as a day-to-day reality; but it made a lot of sense theoretically, as we will see later. Roger Penrose was in fact describing a black hole, where all matter drawn into the hole has to be compressed into a singularity.

The collapse of a star into a singularity which Roger Penrose described was the perfect theoretical description of a black hole (see chapter 11). This artist's impression shows a singularity as a black dot deep inside the black hole. It is so dense that, once caught, nothing can escape the pull of its gravity. Not even light. The three outer light rays are bent around by the gravitational pull of the singularity and the hole it has created; but they each eventually escape. The fourth ends up on the knife edge between escape and falling in and orbits the hole, marking its outer limit. But the fifth, innermost ray gets pulled right in, never to escape.

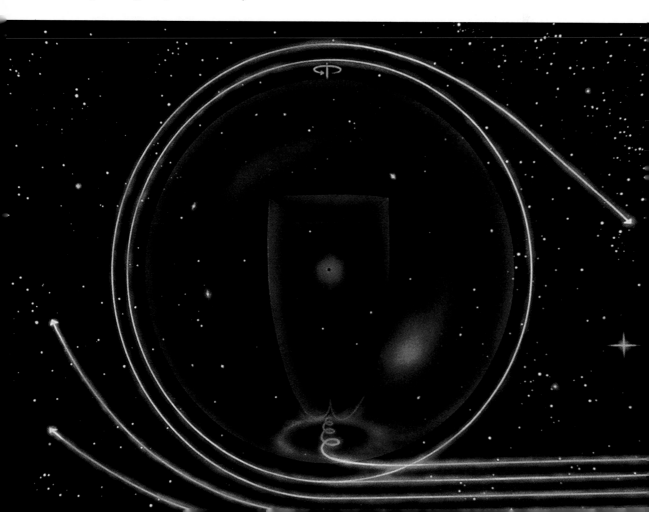

It also gave Stephen Hawking a moment of insight. By reversing the direction of time, and running the event Penrose was describing backward, Stephen realized that he had a perfect model for the Big Bang. A singularity, he argued, was what in Einstein's mathematics corresponded to Lemaître's primeval atom; and it would explode outward with the Big Bang, reversing the dynamics of a black hole and releasing matter as it evolved. Stephen and Roger Penrose published a paper in 1970 which proved mathematically that, if Einstein's mathematics were correct, a singularity had to result from a black hole, and had to exist at the start of the universe. This produced a crisis for physics: how could physics explain everything if its laws did not apply at the very birth of the universe? But, in many physicists' minds, it was also enough to seal the argument for the universe beginning with a Big Bang. After all, the paper argued that if relativity as explained by Einstein is correct—and all the evidence from observation seems to keep confirming it—then the universe must have started with a Big Bang explosion out of a singularity. The equations do not allow an alternative.

Delighted to find that Stephen Hawking seemed to be offering scientific evidence to support the kind of beginning to the universe which one of its own priests, Lemaître, had proposed, the Catholic Church presented Stephen with its own Pontifical Academy medal, in recognition of his work. Perhaps this was a small public sign that the rift between the Church and science opened up by Galileo was slowly being repaired, although the Vatican would perhaps have been embarrassed to discover that Galileo was one of Stephen Hawking's scientific heroes.

Ripples in the cosmos

Stephen Hawking's work on the singularity essentially provided theoretical evidence for the Big Bang. The most recent experimental evidence, and some would say the decisive proof, of Big Bang theory came from a satellite called COBE (an acronym for Cosmic Background Explorer). After Penzias and Wilson discovered the background radiation from the Big Bang in 1965, Hoyle and others had pointed out that it was too even in temperature to have allowed the galaxies to evolve from it. So perhaps it would have to be explained in some other way; if it did not come from the Big Bang after all, it would leave the door open for the survival of "steady state" theory. This provoked some of the supporters of the Big Bang theory to take up the challenge of finding ripples (slight variations in temperature which would allow for the formation of galaxies) in the background radiation.

Without these subtle temperature differences, there would be no reason for any variations in the way the universe had evolved. Everything would have come from a uniform hot soup of energy; when subatomic particles eventually emerged, they would have been evenly spread out. There would have been no areas more dense or less dense than others: just a smooth, even distribution of matter. But the real universe has clearly not evolved like that. Matter has clumped together to form galaxies, with great empty voids between them. In the very early stages of the evolution of the universe, even minute differences in temperature would explain how this happened. Slightly hotter spots would have greater energy than slightly cooler spots; and so more particles could evolve in the hot spots than in the cooler ones. This would allow gravity to draw these denser groups of particles together into tighter and denser clumps, which would eventually also gravitationally attract particles created in the cooler areas. As they were emptied of particles, these cooler areas would evolve into the voids between galaxies, and the hotter areas would evolve into the first primitive galaxies.

George Smoot, a physicist at the University of California at Berkeley, and a big team of experimental cosmologists, who wanted to prove the Big Bang theory was right, were determined to find a way to detect the necessary variations in temperature in the background radiation. They had to believe that the detector originally used by Penzias and Wilson was unable to dis-

Relics, Singularities and Ripples

If the Big Bang started the universe, then certain key things have to be true for the universe to be as we see it today. Within a second of the Big Bang (1) there have to have been minor variations in the temperature of the energy produced. These would only become detectable when the "fog" of the heat has cooled sufficiently to clear (2). This would be 300,000 years later. And from these tiny temperature differences, galaxies and the spaces between them would eventually form (3).

cern these tiny variations. So they knew they would need to develop a refined and sensitive detection system, and also eliminate any possible sources of interference, such as the Earth's atmosphere, which might hide the minute temperature differences in the background radiation.

They tried sending up huge helium balloons, often as big as a football field, in order to lift the complex equipment to the edge of the Earth's atmosphere—equipment which could be as big and heavy as a small car. The balloon was flimsy, no thicker than a plastic bag. It was prey to any changes in wind direction, so the team had no way of knowing exactly what its final observing position would be. The detector could be adjusted to allow for this by remote control and the required data collected, but when explosive bolts were fired to drop the equipment back to Earth, its descent was controlled by a parachute, remotely deployed. But there was no guarantee that it would land in a safe and convenient spot. And any damage to the valuable equipment was obviously expensive.

George Smoot discusses with his team a detail of one of the early detection devices they used. They had to develop a significantly more sensitive system than Penzias and Wilson to be sure of seeing any temperature variations in the background radiation.

For all these reasons, they next went for a more controllable option— a U2 aircraft. A specially indented cockpit cover was used to keep the sensitive detector on the outside of the airplane. Even the window glass of an aircraft would have prevented them taking accurate readings. And in the end, they discovered, so did the movement of the aircraft and the limited time available for taking readings in each patch of sky. The aircraft could not stay still in one position like the balloon; and, even though it could undertake repeated flights past the same spot, it could not get enough readings before it ran out of fuel. The only realistic option, as they had foreseen all along, was to use a satellite. It would operate totally outside the Earth's atmosphere; and it could be controlled by remote firings of small motors to stay exactly where they wanted it in orbit, in a position synchronized with the rotation of the Earth. This would ensure that it offered all the stability of the balloon as an observational platform, and much better working conditions.

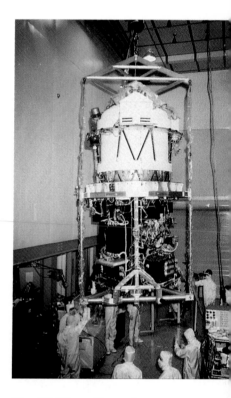

The COBE satellite under construction.

The team knew that winning a place for their experiment with NASA would not be easy, but they prepared their case carefully, and to their delight they were eventually given the opportunity to launch a satellite experiment into orbit. Having built the satellite and tested all their equipment and the remote control systems, they were further delayed by problems within NASA's timetable, brought on by failures in their spacecraft technology. But, once the dramas and crises of the space program were over, they finally got their moment in 1989. COBE was launched on a rocket, and almost immediately began to send back the clearest signals they had yet received. It was soon possible to confirm the background radiation which Penzias and Wilson had observed; but it took another two years to be sure that they had eliminated all possible sources of interference and ambiguity when producing a more detailed computer picture from the COBE observations.

Early in 1992, George Smoot found himself staring at something that really excited him. The computer had generated a map of the early universe from COBE's data which revealed tiny variations in the structure. He was pretty sure about what he was seeing but, just to make absolutely sure, he asked one of his team to analyze COBE's data independently, saying only that he thought a particular approach looked promising. The next morn-

---

The COBE satellite needed sensors which could survive the stresses of a rocket launch into space and still be able to record differences of a fraction of a degree in the temperature of the radiation it was detecting.

ing, Smoot found a computer picture pushed under his office door. It had exactly the same details as Smoot's own picture; and the word "Eureka?" attached to it.

Smoot chose to give the computer model pink and blue colors to distinguish the hotter and cooler parts of the picture. The image quickly became familiar all over the world. The ripples in the cosmos which COBE had discovered were queried and reexamined and carefully checked; but in time their message was widely acknowledged. There clearly were tiny temperature variations in the background radiation from the Big Bang, allowing galaxies to form and evolve into what we see today. Big Bang theory had been spectacularly confirmed.

From Lemaître's inspired idea in 1927 of a primeval atom to Stephen Hawking's own work on relativity and COBE's confirmation of the Big Bang had taken less than 50 years. Not very long when you consider that Ptolemy's model of the universe had lasted as a completely acceptable description of the universe for some fifteen hundred years; and even Newton's infinite unchanging model had survived more than two hundred. So, although the Big Bang theory won wide acceptance during the relatively brief period of those 50 years, it is still not universally accepted even today. After all, it is hard enough to believe that the vast variety of our own planet—its mountains and oceans, its plants and animal life, including ourselves—could all have grown from a singularity, something somehow smaller than an atom and yet immensely dense. And even this hugely complex immense amount of matter is but a

George Smoot with one of the early horn antennas he used in the search for "ripples in the cosmos," which he knew would prove that galaxies could form after the Big Bang.

tiny drop in the ocean; if the Big Bang theory is correct, then it must have spawned, as well as our Earth, all the matter which makes up all the other planets surrounding our sun, which is just one tiny star among billions of stars in one galaxy. One single galaxy among billions of galaxies, all created from a singularity and racing away from each other at colossal speeds as the universe continues to expand some fifteen billion years after it began.

For this is the unavoidable picture you get if you use Hubble's red shift calculations to trace back in time to the point where all the galaxies were in one place: a startling vision of the universe, which this century's advances in physics have breathtakingly put before us. It becomes increasingly difficult to argue against this picture when so many separate pieces of evidence seem to suggest it has to be the right one. Besides Hubble's observations, Einstein's Theory of General Relativity and Stephen Hawking's

The computer generated map of the background radiation which finally revealed the "ripples." Smoot described it as the "cosmic egg" from which everything in the universe could be hatched.

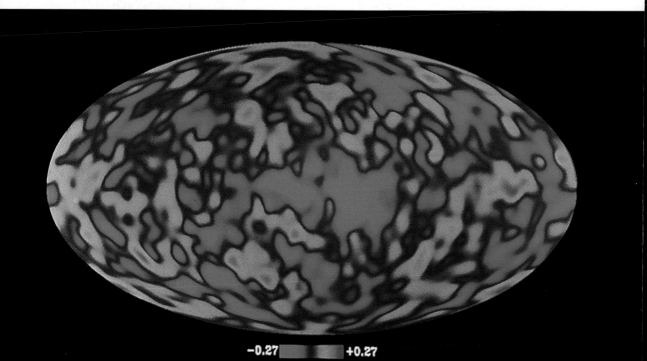

−0.27   +0.27

subsequent work give it huge theoretical support. Then, in turn, the further observational work of Penzias and Wilson, and finally George Smoot, must surely mean we can now be confident that we understand the large scale dynamics of the universe, however unlikely it might seem. And, while the physics of the very large has shown us these extraordinary surprises, the physics of the very small has advanced just as rapidly. This other branch of physics has helped us understand more about the nature of matter, possibly revealing just how the vast amount of matter in the universe could have evolved from virtually nothing at all.

c h a p t e r
s i x

~~~~~~~~

A Matter of Atoms

All things great and small

JUST AS THEY HAD PUZZLED ABOUT THE NATURE OF THE UNIVERSE, THE ancient Greeks also pondered about the nature of everything around them. And unwittingly their thoughts set up a branch of physics which was to grow alongside cosmology and astronomy without ever quite joining up with them. By the twentieth century, our ideas about the nature of the

---

At the center of some galaxies, telescopes have revealed jets of matter being shot out into space for billions of miles. It is extraordinary to think that some of it could end up forming planets like ours, or even our own flesh and blood.

universe had progressed from the mathematics of Pythagoras and Eratosthenes to the mathematics of Einstein: all offering equations and formulas which attempted to define the same enormous relationships between the sun, the stars and the planets. Einstein's physics were principally concerned with the nature of time, space and gravity, rather than the way in which tiny atoms of matter are organized and arranged. The physics that has culminated in Einstein's Theories of Relativity is therefore often referred to as the physics of the very large. But we also need to study the world of the very small—nowadays the realm of particle physics and quantum theory. And the physics of the very small did not develop in anything like the same way as its bigger brother.

The physics of the very large developed in the classical scientific way; ideas about the nature of the universe were confirmed or ruled out by careful observations or scientific experiments. These observations and experiments led in turn to other ideas, which again lived or died as further observations and experiments tested their validity. Ptolemy's view of the universe looked convincing until Galileo's observations made it clearly impossible, leaving the way open for Newton's theories of motion to suggest another model for the universe. Then Doppler's ideas, and von Fraunhofer's discoveries about the nature of light, both experimentally tested and confirmed, allowed Hubble's observations to suggest a model of the universe which began from a single tiny point. Einstein's mathematics showed how the evolution of this kind of universe was consistent with his laws of physics governing space and time; but they did not explain how all the matter in the universe (all the galaxies with their billions of stars, including our sun with its planets) could have developed from one single fundamental beginning—Lemaître's primeval atom.

The idea of some kind of fundamental building block or blocks from which everything else is constructed has a long history. The ancient Greeks, like others before them, thought there were four things that together could forge all the variety of matter we experience around us. They suggested that earth, fire, air and water could always be relied upon to combine in some way to produce everything else. It was obvious that each had special powers: water could dissolve things; fire could heat and melt them; and air could spread fire and dry up water. The earth was the solid stuff which these forces could work on to produce new materials.

The Greeks also felt sure that all matter could be divided and divided again until the tiniest fundamental pieces of the matter being cut up were all that was left. These smallest possible pieces were called atoms—from the Greek word atomos which means "unable to be cut." So, at the time that Pythagoras was trying to give a mathematical harmonious explanation of the relationship between the Earth and the stars, there was a growing conviction that all the substances on Earth would be equally related to each other in a harmonious way, with some fundamental pattern and order to explain them.

Even as late as 1500, illustrations of the nature of everything depicted the four basic elements as Air, Water, Fire and Earth. Here the bird was used to indicate the invisible air, and the earth was implied as the container for everything else.

Sorcerers or scientists?

The search for a fundamental structure for matter did not begin with an intellectual goal; more with an arbitrary determination to see if one kind of matter could be turned into another. The earliest experiments to try to turn one substance into another were not very scientific; they were carried out by alchemists, whose ideas owed as much to superstition and traditional beliefs as science. Alchemy simply attempted to forge precious metals like gold from various base metals by trial and error. Its practitioners essentially assumed that the power of fire, with the right mystical preparations, would somehow transmute one element into another. All they had to do was to keep mixing and heating different things in the right surroundings and they would eventually stumble on the right sequence.

Quite unpalatable things, like dung and urine, were dissolved in water, and gases were evaporated off such mixtures. Solids were smelted into each other, or separated from each other in a molten state. But nothing would give the alchemists the magical transmutation they sought. Increasingly, they turned to astrology and mysticism to discover the secrets of transmutation; and, in doing so, ended up discrediting their fundamental idea that fire, which they saw as energy from the sun and the stars, might somehow be the key to creating one form of matter from another. In fact, as the Church became the center of all learning, it began to condemn alchemy as witchcraft and sorcery, ordering the writings of its practitioners to be locked up or burned.

It is strange to think how close the alchemists came, as we shall see, to identifying the scientific origins of matter, and the source of its infinite variety throughout the universe. They did, after all, believe in the relevance of energy in the sun and the stars. And additionally, despite the condemnation of the Church, alchemy did give science many valuable tools to use in the search for the nature of matter.

Elemental forces

All the alchemists' techniques for separating, dissolving, evaporating and melting materials were taken up by the earliest chemists as they sought to separate out whatever chemicals they could find. They soon realized that many things were compounds of one or more other chemicals, which could be separated and identified by processes similar to those used by alchemists. And they went further; they began to identify chemicals which could not be broken down any more. These seemed like fundamental chemicals; so they were called elements, and were assumed to be the basis of all matter.

Of course the earliest chemists could not have imagined that twentieth-century science would demand an explanation of how all matter could be created from a single source, following the Big Bang. So, theoretically, they had no reason to be concerned if the elements they discovered had no apparent relationship to each other, or if there did not seem to be a way for one element to be transmuted into another. Nevertheless they somehow could not quite believe that everything was to-

This fifteenth-century drawing is thought to suggest that the alchemist it depicts was a charlatan. Even as early as medieval times, alchemy's failure to produce gold had damaged its credibility.

tally disconnected. By the time that they had discovered over sixty different chemical elements, it seemed pretty obvious that the ancient theory of air, fire, water and earth being the basis of everything had to be wrong; but sixty apparently disconnected elements, all of which could be reduced to basic atoms, must have seemed far too many to be fundamental.

For the moment, the idea that the atom was fundamental remained unchallenged. It was just accepted that the smallest possible pieces of each basic element would be atoms of that element. For the early chemists, exploring the exciting relationships between these elements must have seemed much more worthwhile than exploring the nature of the atom. So the pursuit of chemical knowledge centered around finding ways to describe how the basic atom of one element could be distinguished from that of another. The chemists noticed that various chemicals behaved like each other in certain respects. Acids all seemed to dissolve metals for example. Some gases burned easily, while others snuffed out the flame of any lighted taper introduced to them. It was natural to try to group all chemicals together, and in turn the elements, in ways which made sense of the properties they shared. This seemed the most interesting challenge for chemists at the time; and was not at first connected in any way with the work that was going on to define the size and weight, and so the mass, of an atom of each known element.

In fact calculating these details was an extraordinary feat. There was no way to isolate, weigh and measure individual atoms; what was possible, however, was to weigh and measure quantities of chemicals before they were mixed together in a chemical reaction, and then to measure and weigh the volume of the new chemicals produced by the reaction. In this way, some idea of the relative weight and size of different chemicals began to be established: large amounts of some things could weigh very little; small amounts of others were much heavier by comparison. It was soon also realized that it always took a precise amount of heat to raise the temperature of one substance by a specific amount; and a different precise amount of heat to raise the temperature of another chemical by the same amount. A figure was given to express the specific heat, as it was called, of each element. And from this reliable set of figures for the different elements, chemists were able to get a figure to express the atomic mass of each one. Even though you could not actually give figures in terms of fractions of millimeters and milligrams to describe atoms of each element, you could at least give one of the elements a benchmark value, and then describe all the other elements proportionally. At first oxygen was the benchmark; later it was carbon. All that really matters is that a set of figures was developed to indicate the mass, or atomic weight, of each element.

However, when at first these atomic weights seemed to be somewhat arbitrary, and bore no apparent relationship to the similar and dissimilar chemical properties of the sixty-odd known elements, they were assumed to be as insignificant as the number of elements that had been discovered. The chemists accepted that, however many elements there turned out to be, and whatever atomic weight they might happen to have, this would not reveal any connection between one element and the next. Chemical similarities were thought to have nothing to do with atomic weights—until, that is, Dmitri Mendeleev proved otherwise.

Dmitri Mendeleev (1834–1907) was the youngest of fourteen children, and only an indifferent scholar at first. No one would have imagined he was going to make a world-famous breakthrough in chemistry.

Different proportions of chemicals mixed with molten glass create a variety of colors. Perhaps this process led Mendeleev to first think about the importance of atomic weights.

Shuffling the cards

Mendeleev almost certainly developed an interest in chemistry in his childhood; his mother owned a traditional glass factory in Siberia, in which various chemicals were used to color the molten glass in furnaces. It could even be that watching specific amounts of chemicals being carefully weighed out to make subtle differences in the colors eventually led Mendeleev to believe that the atomic weight of different elements might be more important than had been previously realized; but maybe that is too fanciful.

What is certainly true is that the glass factory got burned down; and this led Mendeleev's mother to decide the time had come to put her son's career before everything else. They made the long journey from Siberia to St. Petersburg in 1848 so that Dmitri could become a student at the university there. It is said that the 2,250 kilometer (1,400 mile) journey took them two years to complete; and, almost as soon as they arrived, Mendeleev's

Professor Evgeny Babaev was inspired by the way Mendeleev discovered the Periodic Table.

mother died. But her determination had by then decided her son's career. He earned his keep, as many academics did, by working on government projects; and the task he was given happened to be classifying various petroleum products. It would only have been a short intellectual step from there to imagine that all forms of chemicals could be classified in some way; and so specifically, all the elements.

The story is that he wanted to complete a paper he was writing on the topic one evening, but could not work out how to end it. Frustrated, he went to bed with his task unfinished. Then the answer came to him in a dream. Why not try to classify the elements in terms of their atomic weights? He suddenly felt that this was a much more promising approach than simply grouping them by chemical reactive properties. So he got a pack of playing cards—he was apparently very fond of card games—and began to write out the symbols for all the known elements and their atomic weights—one on each card. Then he began to see what arrangement he could make of them.

The important insight he had, whether in his dream or while arranging the cards on the table, was that the apparently unconnected atomic weights of the elements could be transformed into a meaningful progression if he assumed two things. One was that there were gaps where some elements were yet to be discovered; and the other was that one or two of the known weights could be slightly adjusted to fit. (Because it had not been possible to measure the known atomic weights very precisely, this was not such an unreasonable assumption as it might seem nowadays.) And so it was that, after shuffling and moving the cards about several times, Mendeleev came up with the first correct version of the Periodic Table: the classification of chemical elements into groups within which there are regular steps up in the atomic weight of each chemical.

Mendeleev did not know precisely why there should be such a relationship between the elements; but he was proved right when new elements

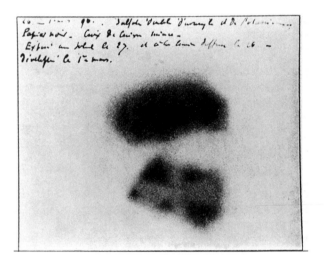

Henri Becquerel (1852–1908) somewhat unwittingly discovered radioactivity. He thought salts of uranium placed on a photographic plate, still carefully wrapped in black paper, would only affect the plate when it was exposed to sunlight. But it was a cloudy day, so he left the plate and the salts in a drawer and waited for better weather. Three days later he developed the plate (*above*) anyway, and discovered it was already fogged by the salts, despite never being unwrapped. His original notes are on the top of the plate.

were discovered which neatly filled some of the gaps he had correctly left in his table. We now know that each step up in the atomic weights of the chemicals is explained by an extra subatomic particle as you progress from one element to the next. The differences between the elements depend on the number of particles it takes to build an atom of each element. But we could not have reached this stage without the insight of Dmitri Mendeleev. He pointed us in the right direction to understand the nature of matter, and how one chemical element is related to another. And, if we are to make sense of everything developing from a single beginning in the Big Bang, there has to be a way in which all chemicals, all matter, can evolve in stages from a common starting point.

A warm blue light

Before he died, Mendeleev is reputed to have said that, if we were to learn more about the nature of matter, we should study the element uranium more closely. His recommendation, if indeed he made it, was uncannily prophetic because at the turn of the nineteenth century, largely in France, the indivisibility of the atom was brought into question because of uranium. Henri Becquerel, among others, had recognized that uranium seemed to give off rays of some kind; and he was determined to find out what this emanation was. As a first step he wanted to see if, as he suspected, it

Henri Becquerel.

was some form of energy. One version of the story suggests that he was initially convinced that sunlight would somehow trigger the energy release, and it could be recorded on a photographic plate (photography had conveniently just been invented). So he wrapped an unexposed photographic plate in thick black paper, and carefully placed a few crystals of a uranium salt on top of the package. But, because the weather turned cloudy, he decided to postpone the experiment. He therefore put everything into a closed drawer and waited for a sunny day to come around. Once it did, and he took out the uranium and the photographic plate, he was astonished to see what had already happened.

More by instinct than anything else, he unwrapped and developed the plate in a photographic darkroom. His intuition was rewarded. A dark shape exactly matching the position of the crystals had been exposed on the photographic plate, just as if the plate had been exposed to light in only that place. Of course the plate had not been exposed to light at all; it was undoubtedly the mark of something emanating from the crystals which, unlike light, had the power to pass straight through the black wrapping paper surrounding the plate.

Michael Faraday had discovered, sixty years earlier, how to generate electricity and in Becquerel's time scientists were beginning to measure all forms of power and energy in terms of the strength of an electric cur-

Marie Curie (1868–1934) learned of Becquerel's discovery and immediately resolved to base her own postgraduate thesis on understanding more about the nature of radioactivity.

rent. So Marie and Pierre Curie, in Paris, decided to set up an experiment to measure the strength of the power which Becquerel had discovered emanating from uranium. They knew that the emanation had somehow been shown to conduct electricity along its path through the air. So they applied an electric charge to a sample of uranium on a plate, to see if it would pass a charge to another plate directly above the sample. Other elements, such as gold and copper, passed no charge to the top plate but uranium always did, suggesting that uranium was indeed the source of an emanation of energy which made air a conductor of electricity.

When Marie and Pierre Curie first discovered radium, it was immediately assumed to be a positive and wonderful thing. No one at first realized there was also a downside to radioactivity.

To measure the strength of the energy emanated, they wanted to set up a repeatable experiment which would show them if the power was always the same, however much uranium they used and however highly they charged the lower plate. There was no way to measure the power directly. But, from Faraday's work, they knew that an electric current could be used to deflect or turn a wire, so they connected the upper plate to a wire. Sure enough—the wire was deflected by the current carried to it from the uranium on the lower plate.

A Matter of Atoms

They then generated an electrical current whose strength they could vary precisely, and fed it to the deflected wire from the opposite direction. This, of course, deflected the wire back in the opposite direction. By measuring how much current it took to deflect the wire back to its starting position, they knew how much power it took to balance the power em-

THE ILLUSTRATED LONDON NEWS.

No. 3642.— VOL. CXXXIV. SATURDAY, FEBRUARY 6, 1909. SIXPENCE.

THE MYSTERIOUS CURE: A PATIENT UNDERGOING THE RADIUM TREATMENT AT THE LONDON HOSPITAL.

In 1909 radium had already been found to have powerful remedial effects; and it was unhesitatingly applied directly, even in purely cosmetic cases. It seemed to have a remarkable ability to remove ugly birthmarks.

anated from the uranium. And so they could calculate the power of the emanation.

It is to the Curies' credit that they conducted their repeated experiments so carefully that they were able to get sufficiently consistent readings of the tiny amount of energy emanating from uranium. But then they discovered something truly extraordinary. The crude source from which they obtained the uranium was pitchblende and they thought they could perhaps save time if, instead of purifying uranium out of it, they used the pitchblende unpurified. They expected to get a weaker current if anything, because of the impurities. But what amazed them was that they got a far stronger current than they had ever measured using pure uranium. There was only one possible explanation. The pitchblende must contain a new element in addition to the uranium, with a far stronger emanation of energy than uranium. The challenge was to isolate it.

The Curies worked hard to reduce crude pitchblende to all its con-

The "wonder" of radium was soon exploited commercially. Luminous watches and clocks became fashionable.

A Matter of Atoms

stituent elements, using all the traditional methods of chemistry—heating, dissolving and so forth. At each stage they then took whatever chemical they had produced and burned it, in order to refract the light from the flame and see what pattern of Fraunhofer lines they could find in the light spectrum. If they saw a completely new pattern, they reasoned, it would have to be the fingerprint of a new element. Eventually they found not one but two radioactive elements, apart from uranium, in the pitchblende. They called the first one polonium, after Madame Curie's native Poland; and the second was radium.

This second one is by far the better remembered. It was certainly important chemically; once the Curies had calculated its atomic weight, it was seen to slot neatly into one of the spaces Mendeleev had left for undiscovered elements in his Periodic Table. But it was also something which captured the public imagination. The emanation from radium was a form of energy easily understood by even the least scientifically minded. The Curies had isolated small amounts of radium so pure that it glowed brightly in the laboratory at night. The beautiful warm blue light was immediately assumed to be something positive and wonderful, providing a healing energy and a decorative loveliness. One dancer at the Folies Bergère is even reputed to have asked the Curies to provide her with a costume covered in radium so that she could dance in the dark!

The more sinister effects of radioactive radium were not immediately recognized. Pierre Curie is known to have suffered burns on his hands from handling the pitchblende; and ultimately the deaths of both Marie and Pierre can arguably be attributed to their exposure to radium. But whichever aspect you dwell on, the crucially important and undeniable fact is that considerable energy was being released from atoms of radium: energy producing enough heat and light to be fascinating to everyone, not just scientists. Now the challenge for science was to explain what was happening inside the atom to release this energy.

Even though understanding the nature of matter was not yet a direct concern of theirs, it was important for cosmologists to resolve this question. The prevailing picture of the universe was still Newton's infinite and eternal model. For atheistic scientists who wanted to reject the idea of a universe created by God, it was vital to explain how all the known elements had naturally evolved in this universe. It seemed the most effective way to counter the arguments of the religious creationists. They claimed that the whole complex variety of everything that existed in the universe could only have been the result of God's wisdom, in making the universe the way we experience it. But, at the time of the Curies, the opponents of the creationists had few real clues as to how matter could evolve. Atoms of the various elements simply seemed to exist as the fundamental building blocks of everything. Mendeleev's Periodic Table suggested that there was a predictable relationship between the atoms of each element; but it offered little insight into how or why all the elements had come into existence.

Discovering that energy could be released from the atoms of some elements suggested that there might be a whole new way to understand the nature of matter. Was there, in fact, something smaller than the basic atom? And, if so, would understanding it help scientists explain how the variety of the universe had evolved naturally, rather than having been, of necessity, created by God? It was time for the beginning of subatomic physics. And it would take only the first few decades of the twentieth century to shatter a belief that had lasted for over two thousand years: the uncuttable atom was about to be proved divisible after all.

~~~

The Energy to Create Everything

Uncomfortably close to alchemy

A BRILLIANT EXPERIMENTAL SCIENTIST FROM NEW ZEALAND, ERNEST Rutherford, who was a friend of the Curies, decided to take their experiments with radium a step further. He worked out an ingenious way to ana-

---

Our Earth seems vast to us, but it is only the tiniest part of all the matter in the universe which was created from the energy of the Big Bang.

lyze what was emanating from the radium and other radioactive elements. The Curies had shown that air could be made to conduct an electric current when the emanation from radium was mixed with it. Rutherford wanted to find out whether the emanation was a stream of energy pure and simple, perhaps mixed with radium in vapor form; or whether it was a totally different chemical which carried the particles of energy. He suspected that, besides the energy which the Curies had measured, there might be another chemical being released, in the form of a gas.

To test his hypothesis, Rutherford constructed two chambers which were linked by a valve that he could open and shut. With the valve closed, he filled one compartment with the emanation from radium, carefully measuring the electrical charge of the gas he was letting into the chamber in the same way as the Curies had measured it. When the charge from the contents of the chamber was equivalent to the charge discovered by the Curies, he knew that he had filled the chamber with the emanation. Then he opened the valve between the two chambers, and measured the charge in the second chamber. He monitored the charges in both chambers, and carefully timed how long it took for the electrical charge in both chambers to become identical. At this point, he reasoned, the emanation from the first chamber would have dispersed equally across both chambers. It was crucially important to know how long this dispersal took, because it was already known that the length of time it takes to disperse a gas will be directly proportional to its atomic weight. So Rutherford expected to get a particular result if radium was being dispersed in vapor form.

But in repeated experiments the dispersal consistently took a totally different time. This clearly indicated that the emanation had an atomic weight below radium in the Periodic Table. In other words, Rutherford had discovered an element lighter than radium, which was being made at the same time as the energy was being given off by radium. For the sake of accuracy, it should be explained that Rutherford did most of the crucial experiments using thorium, which gave off radon gas; and that he got the right result but for the wrong reason! The figures he calculated for the atomic weight of radon were far too low; but they were certainly different enough

Ernest Rutherford (1871–1937) was a New Zealander who did most of his work in Montreal, Canada, and Cambridge, England. He is often acknowledged as the father of nuclear physics; and yet the Nobel Prize he won in 1908 was for chemistry.

The Energy to Create Everything

from the atomic weight of radium to show that a new element was being produced, not simply radium in vapor form. As for improving our understanding of matter, the most significant point was the natural process of change which Rutherford's experiments established was happening spontaneously. One chemical element was undoubtedly changing into another. It was the very process the alchemists had been looking for. As his assistant, Frederick Soddy, is said to have exclaimed: "My goodness, Rutherford, we've discovered transmutation!"

Rutherford was appalled to think that he had done anything which could be remotely connected with the discredited art of alchemy; he refused to countenance calling it transmutation, as if this would somehow cast doubt on the scientific accuracy of his work. But in truth what he had shown was the reality of transmutation; his experiments had revealed that, at least in some cases, one chemical element can somehow be produced from another, releasing energy in the process. It may not have been the alchemists' dream of making gold from base metal, but at least it was a positive confirmation that making one chemical element turn into another was not only possible—it occurred in nature.

As far as cosmology was concerned, if this transmutation process could be shown to apply to all chemical elements, and if everything did start from a Big Bang, then at least theoretically whatever was produced at that time could eventually, stage by stage, turn into everything else in the universe. But proving that the process he had discovered with radioactive elements could apply to all chemical elements was not Rutherford's immediate concern. For him the next challenge was to see what was going on inside an atom which would allow this process of change to take place.

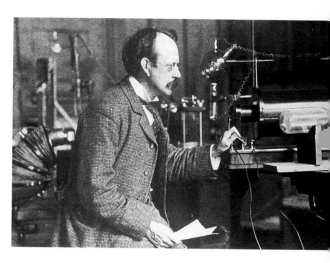

Right Joseph Thomson (1856–1940) was better known as J. J. Thomson. He used a cathode-ray tube (*below*) for his key experiments. German scientists thought the effects in these tubes were caused by electromagnetic radiation before J. J. Thomson proved they were caused by electrons.

The anatomy of the atom

In the early part of the twentieth century, quite a lot of work had already been done on the likely structure of the atom, and several of its components had been identified. The electron, for example, had been discovered by the British physicist J. J. Thomson. Until Thomson's experiments, electricity was known as a flow of energy that could be induced in a suitable metal wire either by chemical means (such as car batteries of the type still in use today) or by physical means (by moving the wire in a magnetic field). It was also known that the current could be made to flow in either direction along the wire; and that, just like the poles of magnets, wires with electricity flowing in opposite directions would attract each other, and wires with electricity flowing in the same direction would repel each other. So the idea of a positive and negative electrical charge representing the two opposite directions was already well established. But no one knew much about what the flow of electricity contained.

Many of Rutherford's experiments were carried out in specially built glass containers. This is the equipment in which he bombarded a gold leaf with alpha particles. *Right* Another piece of apparatus in which Rutherford studied the nature of alpha particles.

J. J. Thomson discovered that it was essentially made up of particles which could be detected when electricity flowed from one end of a specially mounted glass tube to the other. The tube was known as a cathode-ray tube. It was meant to be a vacuum; but in fact some gas remained in it. The tube was held between metal plates which could be given an electric charge: one positive, and the other negative. This allowed electricity to travel from one end of the tube to the other without a wire to join them, and so Thomson could observe what was "in" the electricity without the physical structure of a metal wire to mask it. Because of the presence of the gas, for reasons that were not at first understood, a glowing light was produced in the tube. Several scientists had tried to explain what was going on before but J. J. Thomson was credited with being the first to prove that the cause of the glowing light was a stream of particles which he called electrons. They did not intrinsically "shine" as they traveled along; it was their interaction with the gases in the air in the tube which caused the spectacular glow.

Thomson discovered that the glowing stream of electrons could be influenced by a magnet. And, by observing the direction in which the stream moved, Thomson realized that they had a negative electrical charge. What is more, by showing how far the magnet deflected the electrons, he was able to calculate that each electron had to have a weight less than that of the lightest known chemical atom, hydrogen. So, did this mean it was something more fundamental than the atom? If so, could anything else be found with the right weight and other physical characteristics to make up the constituent parts of an atom? Scientists everywhere began to try to identify what else could be inside the atom along with electrons; and they began to speculate on what kind of structure might be involved. One idea was that the negatively charged electrons would somehow be grouped around a nucleus in each atom, which was positively charged; then the two opposite electrical charges would attract each other to hold the atom together.

Rutherford was able to refine this idea further. Some of his work had involved showing that the energy emanated by radioactive substances took three different forms. In a series of experiments, he found that some of the energy could pass through a thin heavy metal barrier, but some of it could not. In addition, a proportion of the energy which had passed through was held back by a thicker barrier; but the rest of the energy seemed to pass through any thickness of barrier put in front of it. The rays that were reflected back from the thinnest barrier became known as alpha rays; beta rays were the ones which penetrated a given thickness of barriers only; gamma rays were the ones that went through everything.

Rutherford was soon able to establish two things about alpha particles. From the way they were deflected by a magnetic field, he could see first that they were positively charged; and second that they had exactly the mass required for the nucleus of a helium atom. Rutherford

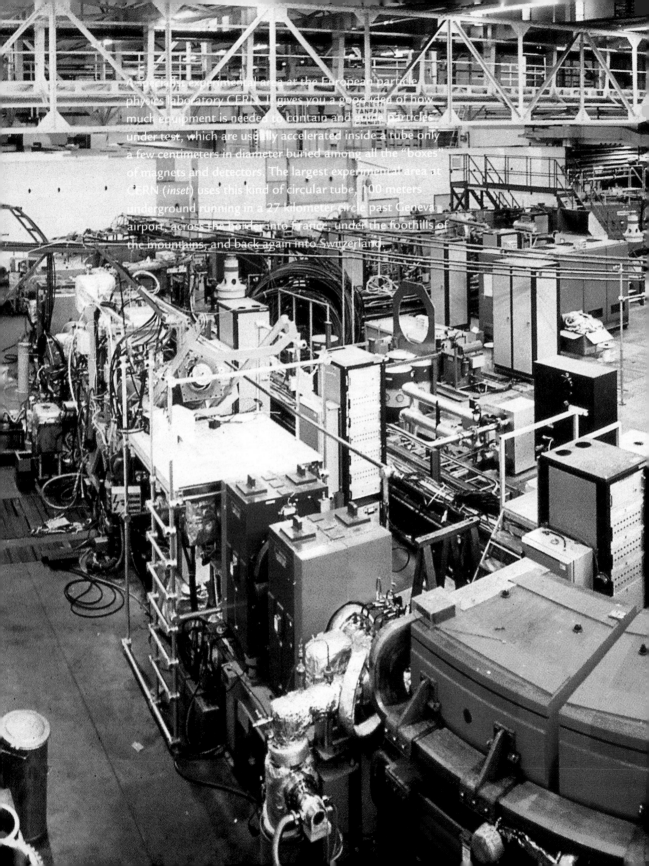

A specialist experimental area at the European particle physics laboratory CERN 1 gives you a good idea of how much equipment is needed to contain and guide particles under test, which are usually accelerated inside a tube only a few centimeters in diameter buried among all the "boxes" of magnets and detectors. The largest experimental area at CERN (*inset*) uses this kind of circular tube, 100 meters underground running in a 27 kilometer circle past Geneva airport, across the border into France, under the foothills of the mountains, and back again into Switzerland.

then tried bombarding a thin piece of gold leaf with alpha particles, expecting to learn something from the way they were bounced back from the barrier. He set up his gold leaf inside a cylindrical container lined with photographically sensitive paper. The paper, once processed, revealed the directions in which the alpha particles had bounced off the gold leaf. Rutherford was thus able to make a microscopic observation of the various directions of the rebounding particles because of the scintillation, or tiny flash of light, each particle produced as it bounced off the gold. This traveled onward to the walls of the cylinder where it registered on the photographic paper. To Rutherford's astonishment, some of the particles seemed to be violently deflected at a huge angle when they bounced off the gold leaf. Rutherford reasoned that this could only be caused by some direct repulsion between the positively charged alpha particle and the positively charged nucleus of the atoms of gold. This implied that there would have to be some sort of space between the electrons and the nucleus, to allow the nucleus to have such an influence. Rutherford proposed that electrons must therefore orbit the nucleus of atoms at a distance away from it.

Accelerating particles

Rutherford himself and the Danish physicist Niels Bohr, among others, eventually worked out an overall structure for the atoms of all the chemical elements; but could they prove what was going on, or learn more about the nature of these subatomic particles? Rutherford's idea for finding out was alarmingly simple; he decided he had to build a machine to split the atom. The machine was known as a particle accelerator; because the idea was to accelerate particles to such a high speed that, when they collided with an atom in a target at the other end of the accelerator, they would split it apart.

By the time the first particle accelerator experiments were being carried out, cosmologists were already familiar with Hubble's observations and Big Bang theory. So they were particularly interested to see if a fundamental sub-atomic structure could be discovered; after all, if the controversial Big Bang theory was going to be believed, there would have to be some way in which the explosion from the tiniest of beginnings could generate all the matter in the universe. If particle accelerators could reveal the true nature of matter, it ought to be possible to see how the basic building blocks might be created at the Big Bang. Alternatively, if they clearly could not be created in such an explosive beginning, then the Big Bang might be shown to be a nonviable theory.

Modern particle accelerators can be huge; the one at CERN in Switzerland includes over 20 kilometers (12.5 miles) of circular tunnel underground extending into France and in parts cutting through the bottom of a mountain. But the principles on which it has been built are essentially the same as those used by Rutherford to construct a glass tube accelerator a couple of meters long. First you need a way to generate the particles; then a way to accelerate them to a high enough speed.

Rutherford's method exploited the properties of electricity. He built something like the positive and negative poles of a battery at either end of a glass tube, to set up an electric field. It is difficult to find an exact analogy; but it is a bit like building a steep hill from one end of the tube to the other. A ball at the top of the hill will naturally run down the hill, gathering speed, until it reaches the bottom at a very high speed indeed. In a particle accelerator, the equivalent of the ball is an electric current, which consists of a stream of electrons introduced at the negative end of the tube. This is in effect the top of the hill; the negatively charged electrons will be repelled by this end of the tube and attracted to the positive end of the tube, the equivalent of the bottom of the hill. And, even over a distance of less than 2 meters (6.5 feet), the electrons will accelerate so much that when they reach the positive end of the tube they will crash with considerable force into a target. The idea is that they will split some of the atoms of the target—usually a simple disc of metal—as they strike them. The result of the collision can be detected on a geiger counter; a device designed by Rutherford and Hans Geiger to detect radioactivity, or the release of energy. So, could the particle accelerator be used to prove that whenever an atom is split it will release energy?

To be absolutely sure that the geiger counter detects only the results of electrons smashing apart atoms of the target, all the air has to be pumped out of the tube, leaving a vacuum; so that there is nothing else, apart from the target, for the stream of electrons to hit. And the results are unequivocal. When an atom is split into subatomic particles, detectable energy is always released. Atoms of the target substance are made to decay artificially in exactly the same sort of way that radioactive chemicals like uranium and radium decay naturally.

The implications of the results are quite clear. Atoms contain energy that is released when they are broken up. But this simply raises more questions. What else, if anything, makes up the subatomic world? Are there any other "ingredients" besides energy that are essential or fundamental to atoms? What is the nature and the role of the electrons and the nucleus of the atom, which seemed to fit so neatly together in the first theoretical models of atoms?

The results from particle accelerators began to offer an extraordinary possible answer. If the electrons and the nucleus of the atom together had the right combined mass to account for the entire atom, then the energy being released had to come somehow from just those particles and the way they are bound together. If, in the natural radioactive decay of radium, a new lighter element is being produced, it must have less mass than the original radium. So, are some of the particles somehow being converted into energy in the radioactive process?

Back to Einstein

As particle accelerators have become more sophisticated, we have been able to develop ways of measuring exactly how much energy is released in a collision; and how much of the mass of the target substance is apparently lost in the collision. The size of the collisions can vary—you can use atoms or particles of different mass for the target—but what is striking is that the amount of energy released is always directly proportional to the loss of mass in the target. This suggests that energy and mass are somehow interchangeable. So at least some of the subatomic particles in a decaying atom must be released as energy.

For scientists, at the time when this was first discovered, it was not such an unexpected idea. Einstein had predicted that mass and energy would be related in his Theory of Special Relativity, in the famous equation $E = mc^2$. He had argued that this equation was an inevitable consequence of the mathematical equations needed in Special Relativity to explain how light always travels at the same speed. His famous equation states that E, energy, is always equivalent to m, mass, multiplied by a constant, the speed of light squared. What this mathematics also predicted was even more extraordinary: that mass would increase, slowly at first but increasingly quickly, as anything traveled at a speed accelerating nearer and nearer to the speed of light.

These are very difficult ideas for most of us to accept. In our experience of daily life there is nothing to suggest that the mass of objects should in any sense be connected with light. But of course we do not experience objects traveling at speeds anywhere near the speed of light; so we are hardly in a position to test Einstein's ideas, improbable as they may sound. But in particle accelerators the streams of electrons used to collide with the targets end up traveling very close to the speed of light as they approach the point of collision; and they can be detected as having measurably increased mass as they accelerate.

The Energy to Create Everything

These results from particle accelerator observations have confirmed, once again, that Einstein's extraordinarily counterintuitive theories were correct; they have proved that mass and energy are totally interchangeable. As the electrons speed up, they gain energy; so the E side of the equation, energy, increases. Since the speed of light is constant, the c^2 part of the equation cannot change. So, for E to equal mc^2, it is the m or mass that has to increase. Which, unlikely as it may sound, is exactly what happens to the electrons in the particle accelerator. More importantly for cosmologists, it suggests that energy is fundamental to all matter. Energy is obviously released when matter breaks up or decays; but can the process be reversed? Can energy be converted into subatomic particles which in turn make up atoms? And could the Big Bang have released enough energy in the first instance to be converted in this way into all the matter in the universe?

Vapor trails

As it happens, we can learn quite a bit more about matter from particle accelerators. In the earliest accelerators it was often enough to register the production of energy with a geiger counter. But an improved way to read the results of collisions had already been developed. In 1895, the British physicist Charles Wilson began working on the development of the first cloud chamber. This was essentially a box full of gas saturated with water vapor. The advantage of it was that it could reveal the pathway of an electrically charged particle which passed through the saturated gas. Although the particle itself is too small to detect visually, its track is big enough to be seen by the naked eye. It is a bit like observing the vapor trail of an aircraft high in the sky; you may not be able to see the airplane itself, but you can be sure it is there because of the telltale trail it leaves. And charged particles make trails in the saturated gas in a similar way to aircraft traveling in a moist atmosphere.

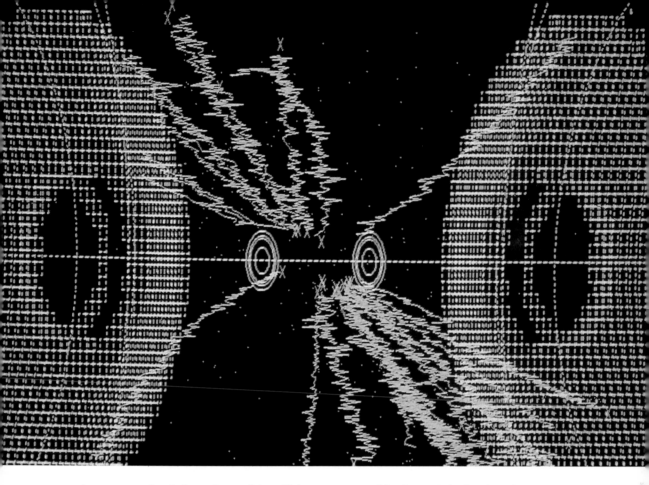

A computer simulation of a particle collision at CERN. This picture is built using the data picked up from a giant detector at the point where particles collide, after being taken to very high speeds around the 27 kilometer accelerator.

Cloud chambers have been used extensively ever since they were invented as a way to detect subatomic particles. Nowadays the electrically charged particle tracks are not literally observed in a cloud chamber; electronic detectors can read the pathways and reproduce them on a computer screen. It is a relatively straightforward matter to have the effects of a collision at the end of a particle accelerator analyzed in this way; and a lot can be learned from the exact pattern of the tracks. The point where the collision takes place is quite easy to see, because typically a great many tracks will suddenly emerge from a single point, indicating the release of a number of charged particles. Usually the paths of the tracks are deliberately detected within a magnetic field, so that the direction in which the track bends will reveal whether the particle is positively or negatively charged. And the amount by which the track is bent will depend on other individual characteristics of the particle, such as its mass. It is therefore possible, over time, to identify the typical shape of tracks made by each individual type of particle. Some are only slightly deflected, for instance, while others can be rapidly spun into an inward spiral pattern. In other words, each particle has a signature track which can be recognized. And when a new, previously unidentified track appears after a collision, its length and the amount it is deflected will give clues as to the mass and characteristic behavior of the particle. In this way we can identify particles predicted by theory when they are revealed in the real world of the particle accelerator.

The search for antimatter

Yet another counterintuitive scientific theory has been proved correct by the characteristic shape of these particle tracks. Paul Dirac, who was Lucasian Pro-

Paul Dirac's prediction of the existence of antimatter was so counterintuitive that many dismissed it as impossible. But within a year, antimatter had been detected.

fessor of Mathematics at Cambridge University (like Stephen Hawking today and of course Isaac Newton some two hundred years before him) showed that, in theory, for every particle there ought to exist a mirror image particle. This was a clear prediction arising out of the equations which explained the way particles had to behave if they were to be able to combine together to create atoms. But it seemed to have an alarming and difficult consequence. If an electron had to have an antielectron mirror image partner (a particle of opposite charge in other words), then the particles making up the nucleus of atoms should also have antiparticle partners. And if negatively charged electrons and a positively charged nucleus made up atoms of matter, what would their antiparticles make? Atoms of antimatter?

The idea of antimatter existing became even more confusing when the theoretical consequences of the existence of both matter and antimatter were worked through. If a particle and its antiparticle were to meet, they would surely annihilate each other in a burst of pure energy. So how could all the matter in the universe exist and be easily detectable if theoretically there should be just as much antimatter around? Either the matter and antimatter should have annihilated each other, or we ought to be able to detect as much antimatter as matter in the universe, which we clearly cannot.

As a first step to discovering a solution to this paradox, it was obviously important to find evidence that Dirac's theory was not totally wrong. Somehow the scientists had to establish that antimatter could really exist. They reasoned that, with so much matter in the universe, any antiparticle would not be able to survive for long without being annihilated by coming into contact with its particle partner. So any antiparticles we could detect would probably come from outer space and be pulled into our atmosphere on Earth, by gravity. As they fell, they would eventually meet a mirror image partner; and the annihilation Dirac's theory had predicted would take place.

As a result, attempts to detect antimatter became quite an adventure.

The story so far . . .

It is, of course, annoying that everything is not sorted out completely, leaving us with a comforting, neat, totally proven picture of the dynamics of the universe. But at least our present understanding of matter fits amazingly well with our present understanding of the way the universe has developed. All the mathematical equations can be fitted together to build an astonishingly precise picture of its evolution.

It all begins with a dramatic Big Bang explosion producing nothing but searing hot energy at first. This energy somehow develops slight variations in its texture as it spreads outward and starts to cool. This allows for slightly hotter spots where, within the first second after the Big Bang, energy

3 seconds   3 minutes

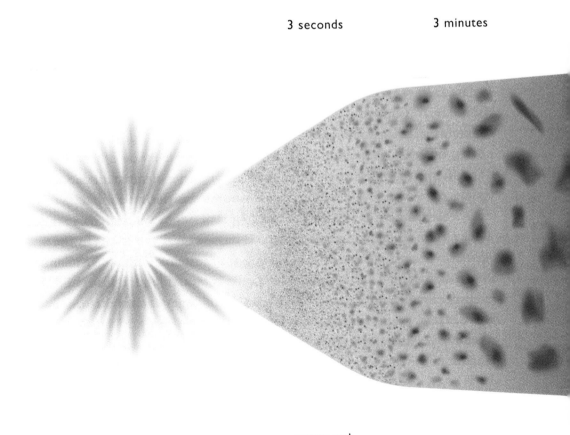

energy and
exotic particles   protons and neutrons

starts converting into particles and antiparticles; and slightly cooler spots which are destined to become the first voids in space. Most of the particles and antiparticles start to be drawn close enough together by gravity for what is known as the electromagnetic force to make them combine; and most of the antimatter is lost in annihilations, leaving only matter swirling in growing irregular clumps. Until three minutes after the Big Bang, it is still too hot

---

No diagram can do justice to the evolution of the universe. It grew from no size at all to a size far greater than our imagination can grasp within minutes. Within a second it had undergone more fundamental changes than it has over the last ten billion years or so. All the matter we now see in all the stars and galaxies was created in the process.

300,000 years   1 billion years   15 billion years

clouds of hydrogen   stars and proto—
and helium atoms   galaxies   the universe today

for these subatomic particles to build anything together; but then some of them start to bind into what will become the nuclei of atoms. It takes 300,000 years for things to cool down enough for electrons to couple with these nuclei to form the first atoms. By then about 20 percent of the nuclei are the heavier type found in helium; the other 80 percent are used to create hydrogen. All the other known chemical elements will evolve much later on.

It takes a billion (1,000,000,000) years, according to all the equations, before millions and millions and millions of these hydrogen and helium atoms have been clumped together by gravity. There are millions of these clumps, each one destined to become some huge cosmic body—typically a whole galaxy. As gravity presses some of the atoms together more and more tightly, the hydrogen atoms begin to fuse in the way Fred Hoyle and his colleagues predicted: stars begin to form within the emerging galaxy and eventually they shine; and the whole life cycle begins, in which all the heavier chemical elements are formed. First hydrogen atoms fuse to produce helium atoms. Then, as the hydrogen starts to get used up, gravitational pressures increase, and the helium atoms start to fuse. One by one, the heavier elements are produced, each in turn fueling fusion reactions to produce the next heaviest element as gravity compresses the star into a denser and denser mass.

Depending on its size, once iron has been formed in the star, it either dies slowly and sheds its elements into space, leaving a white dwarf star, which cools into a brown dwarf (an iron ghost left to wander in space); or the star dies dramatically, exploding in a supernova and creating all the elements heavier than iron in the process. These elements drift through space until they are drawn by gravity into a new heavenly body. If enough matter is pulled in, a new star can be born; but if new fusion reactions do not get started, a planet very like our Earth can result. In the right conditions, life can evolve on such a planet. Which is exactly how we come to be here, observing the marvel of it all for ourselves. After 15 billion years the universe has evolved to be the way we see it today.

There are still many people who refuse to believe such an unlikely sounding story; but, when we piece together all the evidence from the physics of the very large and the physics of the very small, it is hard to construct a better alternative. It is certainly difficult to imagine a more awesome, beautiful or fitting explanation for the immensity and complexity of everything we can see all around us. Cosmology has come a long way from the imaginative application of mathematics to the earliest observations made with the naked eye.

Having embraced the reasons why this amazing evolutionary story has to be true, it might seem that we now have an almost complete picture of everything; but in fact there is still a lot more to be explained. It is suitably humbling to remind ourselves how, first with Galileo's observations, and then with Hubble's, astonishing new evidence made cosmologists revise their ideas quite radically. However well established the current evidence for the expanding evolving universe might seem to be, it is as well to remember that it is largely based on analyzing the light from the proportion of the universe which we can actually see. And there is good reason to believe that there is much more of the universe for us still to detect. In fact what we can actually see may turn out to be less than a tenth of all there is in the universe. Most cosmologists believe that 90 percent of the essential nature of the universe has still to be revealed. It is as if we have only so far looked at the tip of the iceberg.

The Energy to Create Everything

chapter
eight

~~~

Searching in the Dark

An invisible halo

IT WAS, APPARENTLY, ONE OF THOSE NIGHTS WHEN NO ONE WANTS TO GO out. The wind howled angrily as it slashed the rain against the window panes. The cold crept under the doors, reminding you how cosy it was inside the house. But Vera Rubin was determined to make the journey down

---

The Great Andromeda Galaxy is a typical spiral galaxy; but how much of it can we actually see? The central area, packed with millions of stars, may be less than one percent of all the matter in the galaxy. As much as ninety-nine percent could be unseen dark matter.

the freeway to get to the annual meeting of the American Astronomical Association. It was unusual for any member to be given time on the platform to address the complete assembly of members; it was even more unusual for a woman to be accorded the privilege. And Vera knew she had something revolutionary to announce. So, braving the elements, she bundled her young baby into the back of the car, together with her father, who had been persuaded to look after the baby while Vera delivered her paper, and set off into the inhospitable night.

Vera had realized that there was something about the rotation of galaxies which most astronomers and cosmologists had failed to appreciate. If you assumed that all the stars we can see in a galaxy are free to affect each other gravitationally in exactly the same way as gravity governs the dynamics of the solar system, then the galaxy ought to rotate differently. Vera Rubin had checked and rechecked all the recorded observations of certain galaxies, and she was convinced that she had to be right. The galaxies turned like a single giant wheel, not a complex collection of individual stars all in separate free orbits around the heart of the galaxy. As Vera saw it, there could only be one explanation. There had to be part of the galaxy which we were just not seeing and this must make up far more of the total mass of the galaxy than the stars we could observe shining. This dark unseen matter must have so much mass that it held the stars in position, making them the shining hub of a giant invisible wheel. Some stars would be spread out toward and among the dark matter as the spiral arms of the galaxy, rather like cream swirling round on top of a cup of coffee.

Vera presented her paper and was aghast at the reception she got. No one seemed willing to take her ideas seriously. Perhaps there was an element of chauvinism from the largely male audience, still not quite prepared to accept that women could be "serious" astronomers and cosmologists. At least such an appalling prejudice would leave her academic case untouched. But the rejection of her studies was so complete that Vera became seriously disillusioned. Could she somehow have been misled by the observational evi-

Vera Rubin almost gave up cosmology when other physicists refused to believe her ideas about dark matter. But they were the ones who were wrong.

dence? Her confidence was badly shaken. She traveled home in the storm and spent the next few years ignoring cosmology and raising her family.

Of course, ironically, Vera Rubin turned out to be totally right. It is obviously very difficult to prove the existence of something you cannot actually detect; but the existence of dark matter, as predicted by Vera Rubin, has since been confirmed with the aid of computer models. It is possible nowadays to calculate the distances to stars and galaxies pretty accurately, by analyzing the spectra of their light, Cepheid stars and so forth. Repeated observations over a period of time can also reveal any changes in the relative positions of near neighbors to each other, in exactly the same way as the changes in the position of planets in our solar system were first noted. And, by applying either Newton's or Einstein's mathematical equations for gravity to the distances of the planets from the sun, the pattern of movements in the solar system is confirmed. So, if this works for the solar system, why not try to build a computer model of a galaxy using the same principles?

You program the computer with the mathematics of gravity and the distances of all the observed stars from the centre of the galaxy. This enables you to predict the movements of the stars around the heart of the

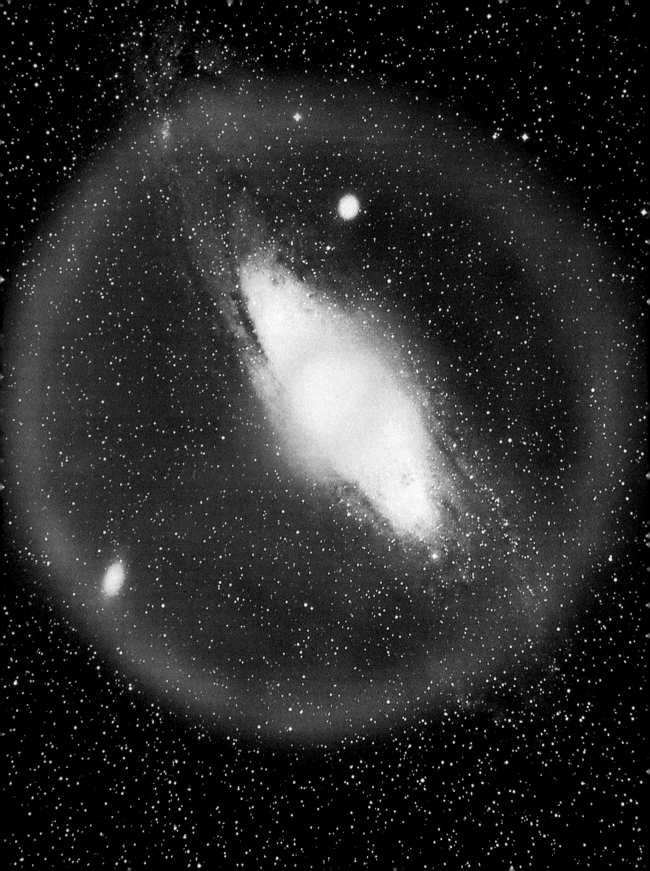

galaxy; and, as a comparison, you use the same mathematics to build a model of the solar system. If this works for the solar system, with nothing else to influence things, you would expect the same formulas to work for the galaxy. But, while the computer will produce a model of the solar system, using these equations, which perfectly predicts the observed movements in the solar system, it will equally clearly fail to build a model of the galaxy which predicts the observed movements in the galaxy. In other words, there has to be something else happening in the galaxy which has not been accounted for. Since there is no evidence of any other forces at work besides gravity in creating the structure of the universe, the only reasonable possibility is that there is dark unseen matter around the galaxy, surrounding it like a huge invisible halo. And, by programming the computer with figures for the nature of such dark matter, it is possible to make the computer model exactly predict the observed movements of the galaxy.

As it happens, the most realistic computer models of the Big Bang also predict a great deal more matter than has yet been observed in the universe; so the existence of dark matter is no longer really in doubt. And all the mathematics point to it making up a staggering 90 percent of all the matter in the universe. But what exactly is it? And what is it like? Is it simply more of the same matter which makes up all the stars we see shining, which has not yet been pulled by gravity into such hotbeds of activity? We certainly find all the elements we now know are made in the stars here on Earth, and hardly any of them emit light in the form in which we find them—only the highly radioactive ones like radium. So it is clear they could

An artist's impression of a dark matter halo around the Andromeda Galaxy. This halo is supposed to be made up of neutrinos; but it could equally contain MACHOs or WIMPs. We only know something must be there because of the gravitational effects on the visible stars in the galaxy.

Searching in the Dark

exist as dark matter. But how could we detect it, millions of miles away, deep in space? At least we know where to look: immediately around the visible fringe of a galaxy where a halo of dark matter is predicted to be. But how can dark matter be identified, if it does not give off any form of heat or light?

The MACHO men

The astrophysicists who first decided to take up the challenge called what they were looking for MACHOs (an acronym for Massive Astrophysical Compact Halo Objects). In order to have the characteristics of the dark matter needed to explain the movement of galaxies, these objects had to have mass so that they would be affected by gravity. They would be found in galactic halos, and would be compact and dense. There was no reason to suppose they would have any of the machismo implied by the acronym; but perhaps it gave an image for the research team to live up to!

They certainly hit upon a bullish approach. Einstein had predicted that light would normally travel in straight lines, as had always been assumed; but

Arthur Stanley Eddington (1882–1944) did a lot of important astronomical work on the nature and structure of stars, and he also understood Einstein's General Theory of Relativity well enough to devise an experimental test of its validity.

he went on to suggest that it would be deflected by the dents in space and time produced by any object with mass which happened to be in its path. He even predicted exactly how much it would be deflected. This was a revolutionary idea for its time, which had taken an extraordinary experiment to prove.

Just after the First World War, the British astronomer Arthur Stanley Eddington led a team which aimed to see if the mass of the sun could deflect the light from a star as the sun passed between the star and the Earth. Naturally, the sun's light would be far too bright for us to see the star when this happened, except in one special circumstance: during a total eclipse. So Eddington and his team carefully worked out the "normal" position of a particular star in relation to its neighbors in the night sky, having determined that the sun would be directly between this star and the Earth at the time of a total eclipse of the sun in the year 1919. The plan was to photograph the area around the sun at the time of the eclipse, when the moon was blocking out all the sun's light. All the stars would thus be revealed in their usual positions; except, of course, for the star being studied.

If Einstein was right, the mass of the sun would have had a gravitational effect on the light from the star, bending it so that the star would appear in a different position from normal. In fact, if Einstein was completely right, it would appear to be in the precise position predicted by Einstein's equations. And, sure enough, when the eclipse came, that was exactly where the star appeared to be.

Eddington's experiment was important because it proved that Einstein's theory of gravity (rather than Newton's) was correct. More specifically, it showed that light was indeed bent, as it traveled through space, by anything with mass which crossed its path. And it was this which caught the imagination of the MACHO men who were looking for a way to identify dark matter. They reasoned that, if some dark matter of significant mass were to cross the light from a star, it would make the star appear to shine

The movement of the stars we can see in the Large Magellanic Cloud—which is only visible from the Southern Hemisphere—suggests that it could have a large dark matter halo rich in MACHOs surrounding it.

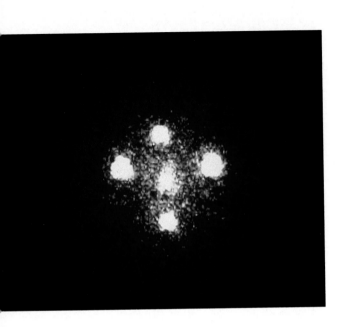

These five similar-looking sources of light from deep in space, as seen by the Hubble Space Telescope, are in fact two objects. The central spot is a relatively nearby galaxy, 400 million light years away; whereas the outer four images are all from a single distant quasar, around 8 billion light years from us. The gravitational effect of the intervening galaxy has bent the light from the quasar, on its way to us on Earth, so that we see the four images of the single quasar. It is a well known example of the "lensing" effect which the MACHO team (*right*) is using.

much more brightly while the dark matter passed by. The reason is logical enough, but not so simple to explain.

Imagine the light spreading out from the star as being made up of a number of individual beams of light. Only one such beam would be traveling in the right direction for us to detect it here on Earth. But, as the dark matter approached first one of the other beams of light, and then the next, it would bend them gravitationally; and eventually one beam would be bent so that it arrived at the surface of the Earth, along with the original, still unbent, beam of light from the star. So the star would appear to shine twice as brightly.

This kind of effect is known as "gravitational lensing," because it is the equivalent of taking a conventional lens and focusing light from a single source on to one concentrated spot. It is just the same as using a magnifying glass to focus the rays of the sun into a single point, perhaps burning a hole in a piece of paper as a result. The important point to grasp, from the MACHO men's point of view, is that this effect ought to take place when-

ever dark matter crosses between a star and us observing it on Earth; and so, if such an event can be identified, it will provide evidence for the existence of dark matter.

It would have to be dark matter concentrated into a pretty large clump, with significant mass and density, if the effect was to be identifiable. And, conveniently, just such large objects were predicted to exist at the end of the life cycle of stars. You will remember that Hoyle and others calculated how evolving stars would keep fusing the next heaviest element in the Periodic Table: first hydrogen, then helium, and in turn all the chemical elements up to iron. The smaller stars would not get beyond the iron stage, because the gravitational forces at work would not be able to generate sufficient inward pressure for the iron to fuse into anything heavier. So the iron star would gradually die, cooling down to leave a brown dwarf star which did not shine—a perfect example of a large piece of dark matter, and exactly the kind of thing which the MACHO men hoped to identify experimentally.

It was no easy task. They chose to do all their observing from the southern hemisphere, in the region of the Large Magellanic Cloud—the site of a small galaxy which was relatively close. They chose it as a place where they could expect there to be a good halo, which ought to include an abundance of dark matter. But the galaxy still contained hundreds of millions of stars. How could they decide which ones to observe nightly to see if they appeared to shine more brightly over a period of a month or two?

The answer was to let a computerized telescope sample a large area of sky, since computerized observation could pinpoint interesting places worth studying more closely. They devised a system whereby a computer monitored the area of sky, having been programmed to identify any single region where there were changes in the intensity of the light being recorded night after night. Once such a spot had been recognized, it could be scrutinized in detail to see what was happening.

And, to their delight, this approach appears to be bringing promising results. There have been a number of sightings which fit the predicted pattern—a star appearing to shine much more brightly than usual over a period of a month or so. Better still, a number of stars have shown this effect in sequence, suggesting a MACHO passing in front of each of them in turn. So we now have some pretty good evidence for the existence of dark matter. However, there are two key questions still to be answered. Firstly, exactly how much of this kind of dark matter is there? And secondly, can it explain the movement of all the galaxies?

The ultimate fate of the universe

Vera Rubin had suggested that 90 percent of the universe would have to be dark matter; in other words, all the stars we can see make up only 10 percent of what is out there. Based on the nature and frequency of their discoveries, the MACHO team are confident that about half of the missing dark matter can be accounted for by the objects they are detecting. But that leaves a lot still to be explained. It suggests that we are a long way from knowing precisely how much dark matter there is in total; or knowing how dense it is, and so how strong a gravitational effect it has. And these are not simply questions of academic interest. The answers will help us discover what is destined to happen to the universe.

The options could not be more staggeringly different. One scenario has the universe ending dramatically; in effect, going into reverse and collapsing to a singularity, imploding back to nonexistence in a Big Crunch. There is no immediate cause for alarm—if it happens, the Big Crunch should not take place for several billion years. Another scenario sees the universe expanding forever, more and more serenely, as the rate of expansion slows down but never quite stops.

To find out which of these sharply contrasting futures awaits our descendants we need to know the mass of the dark matter in the universe. Hubble's red shift observations made it clear that the universe was expanding. The evidence is that the oldest galaxies (the farthest away from us) are traveling away from us faster than the nearer, and so younger, galaxies. As time passes, the spaces between galaxies get bigger. So, when we say the universe is expanding, we mean that space itself is expanding; not that individual galaxies or stars or planets are getting larger. The expansionary force that spreads the universe outward is believed by the majority of cosmologists and astronomers today to be the explosive outward force generated by the Big Bang. Whatever it is, it is there. And, what is more, it can be seen to be diminishing in speed as time goes by.

This conclusion comes from observing galaxies on a regular basis and checking their speed from time to time, by seeing how far the Fraunhofer lines are shifted towards the red in their light spectra. It turns out that the speed at which they travel away from us is getting progressively slower as time passes. This means that there has to be a force countering the overall expansionary force and slowing it down; otherwise it would simply carry on unchecked for ever. And the most obvious counterforce we know of is gravity. According to both Newton's and Einstein's theories, the more matter there is in the universe, the stronger the overall pull of gravity inward, trying to clump all this matter together in one great central lump. So far, of course, the universe has not only stayed spread out; it also continues to expand as we observe it today. So, at the moment, the expansionary force is slightly more powerful than the gravitational force opposing it.

But the expansionary force is also slowing down somewhat. This suggests that there must be enough matter in the universe to have this slowing-down gravitational effect; but it does not tell us whether there is enough matter eventually to slow down the expansionary force to a halt, and then begin to pull the universe together into a smaller and smaller, increasingly dense clump which would eventually implode in the Big Crunch. Alterna-

Although they will all admit that we simply don't have enough data yet to know the answer, cosmologists like to speculate on what kind of an end the universe will have.

We're running out of time

I once thought I would have liked to have written a book called *Yesterday's tomorrow: a history of the future.* It would have been a history of predictions of the future, nearly all of which have been very wide of the mark. But I doubt if it would have sold as well as my history of the past.

Foretelling the future in antiquity was the job of oracles. These were often women who would be put into a trance by some... the fumes. Th... be int... roundi... lay in... The... Greec... hedg... ous... wha... Per... ora... wi... wi... of... p...

Is the universe going to expand into eternity or will everything collapse in one Big Crunch in which physical laws become meaningless? **Stephen Hawking**, author of the phenomenal bestseller *A Brief History of Time*, sheds light on the darkest regions of space and time and considers an extraordinary array of possibilities for our future

If the dens... one second... had been... in a tha... universe... apsed a... other h... less u... univ... ess... ab... it...

This means that even if there... events after the Bi...

Universe 'to end in Big Crunch'

by Robert Matthews

PROFESSO...

...wk...
...und
...hat
...lly
...in
...e-
...a-
...s

he has found a way of getting around this problem: the "no boundary condition".

He explains his idea by analogy with a globe. Its surface is limited but, paradoxically, an insect could walk all over the globe forever without meeting an edge. In mathematical jargon, the globe's surface is "finite but unbounded".

Prof Hawking has shown that it is possible to think of space and time as being like the surface of a globe — also finite yet having no beginning or end.

The "no boundary" condition can thus get around the problems surrounding the moment of birth of the universe. But according to Prof Hawking, it also makes a prediction about how the universe will end: in a collapse towards a cataclysmic Big Crunch.

His la...

Are we masters of our fate?

In the play *Julius Caesar*, Cassius tells Brutus, "men at some times are masters of their fate". But are we really masters of our fate? Or is everything we do determined and preordained? The arguments of preordination used to be that God was omnipotent and outside time, so God would know what was going to happen. But how then could we have any free will? And if we don't have free will, how can we be responsible for our actions? It can hardly be one's fault if one has been preordained to rob a bank. So why should one be punished for it?

In recent times, the argument for determinism has been based on science. It seems that there are well-defined laws that govern how the universe and everything in it develops in time. Although we have not yet found the exact form of all these laws, we already know enough to deter-

STEPHEN HAWKING

BLACK HOLES
AND BABY UNIVERSES
and other essays

Arguably the most brilliant scientist of his generation, **Stephen Hawking**, in the first of two essays from his latest book, asks: how can the human race have free will if the universe is determined by the laws of science?

A third problem with the idea that everything is determined is that we feel that we have free will — that we have the freedom to choose whether to do something. But if everything is determined by the laws of science, then free will

form of life, whatever that may have been. As evolution progressed, it would have led to the development of the central nervous system. Creatures that correctly recognised the implications of data gathered by their sense organs and took appropriate action would be more likely to survive and reproduce. The human race has carried this to another stage. We are very similar to higher apes, both in our bodies and in our DNA; but a slight variation in our DNA has enabled us to develop language. This has meant that we can hand down information and accumulated experience from generation to generation, in spoken and eventually in written form. The effect has been a dramatic speed-up of evolution.

It took more than three billion years to evolve up to the human race. But over the last 10,000 years

would differ on details like the distribution of stars and, even more, on what was on the covers of their magazines. (That is, if those histories contained magazines.) Thus the complexity of the universe around us and its details arose

tively, there might not be quite enough matter for gravity to dominate the expansionary force, in which case the expansion would never quite stop, but would continue forever, always getting slower and slower.

It is difficult even to take an educated guess about what the final answer might be. We are obviously still a long way from knowing how much mass there is in the universe; the amount detected as MACHOs does not even suggest enough dark matter to account for the rotational movement of galaxies. So, what other dark matter might there be?

One promising-sounding clue comes from the theories of subatomic physics. We have already observed in particle accelerators how energy can become particles which in turn become energy again, performing an exotic dance routine as energy and matter try to settle into more stable forms in the most extreme conditions. Some of the particles produced during this process are indeed exotic: particles which have strange properties, not easily detectable experimentally. But their existence is essential if we are to make sense of the equations explaining violent nuclear reactions. The observed consequences of huge explosive collisions of particles can only be predicted by mathematical models if they also predict the production of these other, more exotic particles. And, since the Big Bang must have been the most extreme nuclear reaction of all, it should have produced an abundance of exotic particles. It makes sense to try to detect them—to see if they can explain the missing dark matter, and possibly the ultimate fate of our universe.

chapter
nine

~~~~~~~

Exotic Excursions

SEARCHING FOR ORDINARY DARK MATTER IS DIFFICULT ENOUGH; THE MACHO men can attest to that. It is hardly easy to find one star in millions which appears to increase in brightness for about a month while dark matter passes by. But searching for exotic dark matter is even more of a challenge. Cosmologists know, from theoretical calculations, the characteristics of the particles they want to find; and these characteristics nearly always ensure that the particle is going to be hard to detect.

The Trifid Nebula is a huge cloud of mainly hydrogen gas some 3,000 light years away. It glows red because of the effect of ultraviolet radiation from young stars in its midst. Clouds like this raise the likelihood that there are other such huge clouds of dust in the universe where stars are not shining; unseen clouds of dark matter.

Fishing for neutrinos

An obvious example is the neutrino which is certainly a particle worth hunting for. According to the equations, every nuclear reaction ought to produce an abundance of them. But they have to be so tiny and so unlikely to interact with other things, that they will pass through anything in their path—our own bodies, the theory says, are being constantly bombarded by neutrinos from the nuclear fusion reactions going on in the sun. These elusive little particles apparently pass straight through us, quite untroubled, and continue their journey, passing into the Earth and traveling on out of the opposite side of the planet! So what chance has a detector of actually picking one up? Especially when, according to the theory, a neutrino has neither an electrical charge nor any mass (two of the essential characteristics which have traditionally helped physicists detect particles in cloud chambers and particle accelerators).

You might well wonder why it is even worth bothering to try to detect them if they have no mass. After all, without mass, however many neutrinos we can detect, they will make no difference to the overall force of gravity in the universe; in which case they cannot be part of the missing dark matter. However, if they turn out to have even the tiniest amount of mass, so many of these neutrinos are thought to exist in the universe that they could explain most, if not all, of the missing dark matter.

The existence of neutrinos was predicted in

Wolfgang Pauli (1900–1959) an Austrian physicist whose work on the laws governing subatomic particles won him the Nobel Prize for Physics in 1945. He first proposed the existence of the neutrino in 1930. Neutrinos result from what is known as beta decay in nuclear reactions, such as the explosion of an atomic bomb.

1930 by Wolfgang Pauli—a neat solution to the problem of what happens to energy in nuclear reactions. It took another twenty-five years before anyone was able to detect one; and a further ten years before anyone found one that had to have occurred naturally in some part of the evolution of the universe. An American, Frederick Reines, was central to both discoveries. He had, in his younger days at least, a reputation for being unable to resist an impossible challenge. And he apparently wanted to track down a neutrino

The bomb dropped on Hiroshima in the year Pauli won his Nobel Prize would have been a rich source of neutrinos.

Fred Reines worked on the U.S. Armed Forces Special Weapons Project between 1949 and 1953, and would have had access to bomb tests, where he hoped to detect neutrinos.

as much for that reason as any other. His first idea was certainly impossible. He reasoned that, if neutrinos were a product of nuclear reactions, the logical place to look for them was at the heart of a nuclear explosion! He seriously considered trying to build a detector which would survive the blast of an atomic bomb test explosion. But the instruments could not be both delicate enough to detect the neutrinos and tough enough to remain undisturbed by the explosion.

What he did do in the 1950s was almost as crazy. He worked out that a nuclear power station, even though its nuclear reactions had nothing like the force of a nuclear bomb, would produce a large enough number of neutrinos to make it reasonably likely that they could be detected nearby. After all, they would pass through the shielding on the reactor, which contained the other subatomic particles resulting from the nuclear reactions, and travel on out into the neighboring countryside. But knowing the neutrinos were going to be there was one thing; constructing a detector capable of revealing such a particle was quite another. After all, this was a particle which could pass through a shield around a nuclear reactor, had no electrical charge, and precious little, if any, mass.

The answer, of course, was not to look for the particle at all, but to look for the results of its passing by. When even the most insignificant particle hits something, there is a tiny release of energy, detectable as the briefest and most minuscule flash of light—the sort of thing Rutherford had observed in his gold leaf bombardment experiments which had helped determine the structure of the atom. Neutrinos would be no exception to this rule so at least, whenever they hit something, there was going to be a sign of their presence which could theoretically be detected. It is also relatively simple to build a detector which will record the release of tiny bursts of light. The problem was how to distinguish when this scintillation of light, as it is called, was produced by a neutrino rather than anything else. In an unmodified environment there would be thousands, if not millions, of scintillations detected for every one which was caused by a neutrino. Separating the neutrino events from all the others would be like looking for a needle in a haystack.

But Frederick Reines loved a challenge. He worked out how to make a viable neutrino detector by patiently eliminating as many non-neutrino events as he could, so that the possible candidates for neutrino-induced scintillations could be reduced to a more manageable number. Each one could then be studied closely to see if it was consistent with neutrino theory. First of all he had to find a suitable site for his experiment. He arranged for it to be fairly deep underground, where a number of possible sources of interfering particles would be naturally eliminated because they could not penetrate that deep below the surface of the Earth. Next, he shielded the detector even more carefully than they had shielded the nuclear reactor nearby. But, whereas the nuclear power engineers wanted to build something capable of keeping most subatomic particles trapped within the shield, Reines and his colleagues wanted something that would keep as many particles as possible outside their shield. That way, only the most penetrative particles, like neutrinos, could possibly reach the detector and produce scintillations.

Having completed all these preparations, they did a series of calibrating experiments in which they deliberately fed an electronic signal to the detector. This was carefully designed to mimic the behavior of the particle they were looking for: in this case, of course, the neutrino. It is incredible how accurately it is possible to do this. It means computing the characteristics of the neutrino into the electronic signal so that its impact on the detector can be confidently believed to be identical to a true neutrino impact. The pattern produced by the detector then provides a portrait of what will happen if a true neutrino strikes the detector when the experiment is up and running. Once the calibration is over, and several electronically produced artificial neutrinos have been observed to check that they will all produce the same characteristic portrait of a neutrino event, it is simply a matter of waiting. It's a bit like fishing: once the hook, line and sinker have been properly prepared, it is just a matter of sitting on the bank and hoping that something will come along.

When a likely candidate event is detected, all the possible data about that event is recorded on a computer and carefully analyzed. Only when all other explanations have been tested and clearly found wanting will the team accept that they have indeed seen a neutrino event. This stage of the process is rather like solving the murder in a detective story. One by one, all the suspects are eliminated until there is only one possible suspect left; and you then know that you have discovered the murderer. In the same way, it is only when the neutrino is the only possible explanation of what the detector has recorded that you allow yourself to say that you have seen a neutrino. And Frederick Reines and his colleagues eventually satisfied themselves that they had done this. In fact, in their underground laboratory, they were soon detecting about three neutrinos an hour.

Do neutrinos have mass?

As far as cosmology is concerned, establishing that neutrinos actually exist is only the first step. Knowing they are produced in man-made nuclear reactions on Earth does not prove that they are naturally produced in nuclear reactions in stars or at the Big Bang. It certainly makes it more likely, but

Yves Declais outside the nuclear power station in northern France where he hopes to discover if the neutrino has mass.

the cosmological significance of these particles could only be more soundly based when neutrinos had been detected reaching Earth from space. And neutrinos could only be a possible candidate for part of the dark matter in the universe once it had been established that they did indeed have some mass, however little it might turn out to be.

Frederick Reines could not find out about the mass of the neutrino; but he did manage to confirm that this particle was produced naturally in the evolution of the universe. Typically, there was even a touch of swashbuckling adventure about the way he did this. He needed to take his detector and all its shielding down as deep as he could below the surface of the Earth, as far as possible from any nuclear power plant or atomic bomb test site. He also needed to be somewhere where the local geology could not be responsible for the release of neutrinos, in order to be sure that anything he detected could only have come from outer space. So it was that he ended up down a gold mine in South Africa, and in 1965 he detected something at least as rarely discovered if not as precious as gold; a neutrino brought to Earth amid cosmic rays.

Others have continued to work on the problem of the mass of the neutrino; and one group in northern France are hopeful of success in the

Exotic Excursions

near future. Yves Declais and his team, housed in a renovated chateau near a nuclear power station, have set up a stylish and ingenious experiment worthy of Frederick Reines himself. They are about to carry out the second half of a project which began in Belgium, near another nuclear power station. There they effectively repeated Reines's experiment to detect neutrinos, but in addition they carefully measured how far they were from the heart of the power station's reactor. Their aim was to see how many neutrinos they could detect every hour at this specific distance from a nuclear reactor. Only when these figures were carefully recorded, and a predictable pattern established, were they ready to move to the French power station. Here they have been able to set up exactly the same experiment as in Belgium; only this time at a precisely measured greater distance from the nuclear reactor. At the moment they are still calibrating their equipment, and carefully noting any local differences caused by the change of locale. Then they want to see whether there is any effective drop in the overall rate at which neutrinos are detected this much farther away from a nuclear reactor.

The thinking is that any statistically significant difference can only be accounted for if some of the neutrinos are decaying: changing into other types of particles and releasing energy, much as radioactive elements like radium naturally do. One of the requirements for a substance to decay in this manner is that it must have mass. So, if the neutrinos can be proved to be decaying, it will mean they have mass. And, even though it will take an even more ingenious experiment to determine exactly how much that mass is, at least we will then know that neutrinos occur naturally in the universe and

Particle tracks in a bubble chamber showing a neutrino interaction event. Coming in from the bottom of the picture, the neutrino, which produces all the other sprays of particles when it interacts with a proton, has no electrical charge and so it cannot be seen—but the aftermath of the collision gives it away.

also, that because they have mass, they can account for some, if not all, of the missing dark matter in the universe.

Meanwhile, theorists have found a way to examine the implications if neutrinos do turn out to have mass. This uses the same computer modeling techniques which confirmed Vera Rubin's idea that there had to be dark matter in the universe, to explain the rotation of galaxies. These techniques have rapidly become more and more sophisticated. Carlos Frenk, for exam-

Typical computer pictures of the structure of the universe. The yellow image was built up from observational data of the actual universe; the green image has been modeled by the computer assuming that the dark matter is hot (for instance, neutrinos). The blue image, using cold dark matter, is clearly much more like the (yellow) actual observed universe.

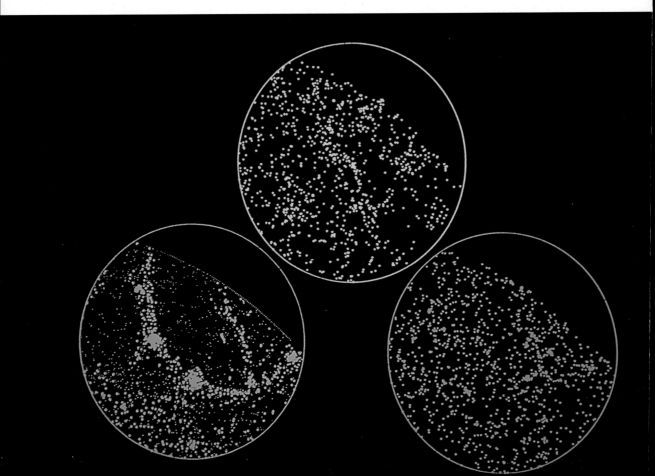

Carlos Frenk has built up similar computer images, revealing how the galaxies are distributed.

ple, a Mexican cosmologist with a German father, has used them to build models of the whole universe. He truly represents the increasingly global collaboration between scientists trying to unravel the final secrets of the universe; he is now a professor at Durham University, having met and married a Scottish undergraduate who was studying Spanish while they were both at Cambridge University!

He has built a database containing all the facts we know about the universe—such as the rate at which it is currently expanding, the size and mass of observed galaxies, their present distance apart from each other, and so forth. This gives him some of the essential information for building a computer model of the universe, which he can run backward and forward in time. To complete his computer programs, he adds the equations that should explain the dynamics of the universe. These include Einstein's General Theory of Relativity and the ways in which particles and energy are predicted to behave (most of which has been confirmed in particle accelerators). He then asks the computer to build a universe based on the information he has programmed into it. Obviously, if all the necessary ingredients were in place, you would expect his computer to produce, 15 billion years or so after the Big Bang, a model that closely resembled the universe we see today.

The first model he built included only the amount of matter we can actually see in the universe. As you might expect, it ended up looking nothing like the present universe. There was simply not enough matter to bring about the gravitational effects which would let the galaxies evolve. Instead,

Exotic Excursions

it left a universe of thinly spread clouds of matter—a bit like an uneven but all-encompassing fog.

This seemed to confirm that there must be dark matter out there, still to be discovered, which fundamentally affects the way the real universe actually works. So Carlos and his team then set about building another model: this time, one in which neutrinos were assumed to have mass, and would make up the missing dark matter. Perhaps now we would see something more like our known universe. To make sure everything was properly allowed for, the model was put together very carefully, over several months. Finally it was ready to run.

Carlos and his colleagues were breathless with excitement as they watched the picture of their model universe emerge: yes, there were definitely galaxies forming now. But then the excitement suddenly died. The pattern of the galaxies which were forming bore no resemblance to the pattern of galaxies we observe in the real universe. Even with mass, the fast-moving neutrinos could not pull galaxies close enough together to match the observed reality. The neutrinos, on their own, were simply not able to create the gravitational effects which would prove that they made up all of the missing dark matter. There had to be something else—another exotic particle yet to be discovered.

Looking for WIMPs

Carlos next programmed his computer to describe the characteristics of a particle which, if it existed, could adjust his model to match the way we see the real universe. The answer he got was not too surprising. It would need to be something much less mobile than the neutrino; a particle with a lot of mass which would just sit there, not interacting much with other particles

Neil Spooner set up his experiment in Boulby mine to find WIMPs after his father, a mining engineer, advised him that it was the deepest mine in Britain.

but simply creating enough gravitational pull to hold the galaxies much closer together. It was soon dubbed "cold" dark matter, meaning that it was slow and inactive by comparison with the fast-moving neutrinos, now known as examples of "hot" dark matter.

But perhaps the most exciting thing about the particles predicted by the computer was that their detailed characteristics uncannily matched those of a particle whose existence had been predicted by a totally separate branch of physics, quantum mechanics (the physics of subatomic particles and the interactions between them needed to make up atoms). The dynamics of the building blocks of matter predicted by quantum mechanics would only work if there was an undiscovered particle involved, a particle exactly like the one the cosmologists' computer had come up with. There was now rapidly growing interest in trying to discover whether such a particle existed.

In cosmology, where MACHOs had already been given their acronym to describe one sort of dark matter, someone wittily described the new particles as WIMPs (standing for Weakly Interacting Massive Particles). And being able to interact only weakly with anything meant that detecting them was going to be extremely difficult. After all, if they rarely interacted with anything else, there would not be many occasions on which they would reveal their presence. The neutrinos had posed a similar problem. So, not surprisingly, the detectors set up to search for WIMPs were very similar to those used by Frederick Reines. The bottom of a deep mine was the ideal site; and well-shielded detectors were again a must. One particularly useful mine is in Yorkshire in northern England. It is extremely deep and principally mines potash. This means that it is geologically about as good as you could hope for, providing a very high probability that there will be no locally produced events to confuse the detectors. So a team from Sheffield University have set up an experiment about 1.6 kilometers (1 mile) below the surface of the Earth; and, like Frederick Reines and Yves Declais, they have patiently sat and waited for something to happen. But they have not, as yet, been as well rewarded as the neutrino teams. So far, no WIMPs have shown up—here or anywhere else in the world.

The research teams are not deterred. They reason that, because of their weak interactions, we cannot expect to detect WIMPs easily anyway. In their minds it is only a matter of time; all that is necessary, they believe, is patiently to increase the sensitivity of their detectors until the right events can be detected. This makes life difficult because the more delicate the detector, the more likely it is to be damaged by pressure changes as they take it down the mine. The Sheffield team has already lost at least two expensive pieces of equipment this way.

But they have received encouragement in their patient vigil from recent observations by a team of astronomers. Sandra Faber, an American as-

The Hubble telescope took this amazing true-color picture. The features which make it remarkable are not the larger or more yellow spots (these are foreground stars and galaxies) but the tiny blue spots which are directly observed galaxies some 8 billion light years away, and so are being seen as they were very early in the life of the universe. The blue color comes from intensive star formation. Although it must be there, there are no obvious clues to reveal the presence of dark matter.

Sandra Faber, Professor of Astronomy and Astrophysics at the University of California, Santa Cruz, is mapping out major galactic movements which she is convinced have to be caused by dark matter.

tronomer who lives in San Jose, California, has—together with her colleagues in a collaboration somewhat vividly known as "the seven samurai"—refined the techniques for analyzing the spectra from starlight. Her team can now get a three-dimensional picture of how stars and galaxies are moving in relation to each other. This was not the main aim of their work originally, but it has allowed them to draw up a three-dimensional map of much of the universe. And when this was developed into a dynamic computer model, it revealed some astonishing movements.

Although the overall picture is of all the galaxies steadily moving farther away from each other, in large parts of the universe there appear to be strong countercurrents of movement, often involving whole clusters of galaxies. One massive group of galaxies, for instance, including our own Milky Way, seems to be hurtling at great speed towards an area which has consequently been dubbed "the great attractor." This suggests that there have to be even more gravitational effects in the universe, totally independent of the halos around galaxies, and caused by dark matter with the characteristics of WIMPs. This dark matter causes even more effects which neither MACHOs nor neutrinos seem able to explain on their own.

All this also underlines just how much of the universe remains a mystery. Despite being able to trace back in time to within a second of the Big Bang, cosmologists cannot yet solve all the mysteries of dark matter. Nor can they make any confident predictions about the ultimate fate of the universe. Will it end in a reverse of the Big Bang, with the entire universe collapsing slowly backward until it is finally compressed into a singularity at the Big Crunch? Or will it continue expanding forever, with the expansion slowing down all the time but never quite stopping? It all depends on unravelling the mystery of dark matter, and perhaps some of the other improbable-sounding realities which the universe seems to delight in revealing.

∿

Extraterrestrial Intelligence and the Baffling Quasars

Tuning in to the universe

IN THE 1950S, A NEW WAY OF STUDYING THE UNIVERSE WAS BEGINNING
to make its presence felt. Radio telescopes had been around since the late
1930s; but their resolution—their ability to distinguish the precise wave-

---

When radio telescopes started to reveal extraordinary signals
from far across the universe, they triggered an almost
unbelievable chain of cosmological activity which was to end up
with the discovery of black holes.

lengths of far-off signals—was dramatically improved in the 1950s. And the result was that radio signals began to be detected from all sorts of unexpected places in space. Some people's imaginations conjured up the most fantastic possibility: since we send out signals on radio waves ourselves, could these signals be broadcasts like our own—messages from other intelligent beings?

As it happens, this is not such a wild flight of fancy as it might sound. Intelligent life obviously evolved on our planet; and the likelihood of the Earth being unique among planets has to be pretty slim. After all, ours is only one of nine planets orbiting one star, our sun (one star out of millions in our galaxy, which is in turn one galaxy in millions across the universe). Out of billions and billions of stars, is it really likely that only our sun has a planet on which intelligent life has evolved? With so many stars, surely at least some of them must have planets in orbit around them, just like our

Radio telescopes used in arrays, like these in Australia, can combine data to acquire very high resolution images which appear as if a single telescope dish with a 3.5 mile diameter has collected the data.

Optical telescopes are too delicate to be constantly exposed to the atmosphere. A domed roof is rotated to line up with whatever is to be observed, and then opened to allow observation to take place.

---

sun; and some of those planets must have the ideal conditions for the evolution of intelligent life. We have known for a long time that our sun is just another star in just another galaxy, in no particularly special position in the universe. Why should there not be other parts of the universe where life has evolved on another planet exactly like our own?

Extraterrestrial Intelligence and the Baffling Quasars

The idea that there might be other intelligences trying to communicate with us from elsewhere in the universe was certainly taken seriously enough for the possibility to be scientifically investigated. The SETI (Search for Extraterrestrial Intelligence) Institute was founded, and funded, at least initially, by American government money. It undertook to search for extraterrestrial intelligence by analyzing radio signals from other parts of the universe, to see if any of them had patterns which suggested they were deliberately generated, rather than totally arbitrary and random emissions as a result of natural occurrences. The work still continues today; and frustratingly, despite technological improvements in the way signals are sampled and analyzed, so far there has been no success. In the end the American government lost faith, and the SETI Institute now survives only thanks to money from computer industry companies. But it has at least confirmed something important for science. A huge number of signals all across the universe have been found to come from some natural phenomenon rather than a purpose-built radio transmitter.

The mystery of the quasars

From the moment radio telescopes began detecting these signals, it was natural for astronomers to train their optical telescopes on the places where strong radio signals had been picked up. Once the light had been analyzed, they usually saw quite dramatic but not totally unexpected phenomena which explained the radio waves. It had already been confirmed that the force of gravity sometimes brings stars or even whole galaxies so close to each other that they merge into one, often involving quite violent collisions. And when optical telescopes were used to look at the places where the radio telescopes had picked up signs of radio emissions, quite often they saw something like two galaxies in collision, producing a high level of nuclear

Maarten Schmidt, at the California Institute of Technology, discovered the nature of quasars by realizing just how far their light was red shifted.

reactions and consequent emissions of radio waves. But not all the sources of radio waves turned out this way. In fact a few looked much less dramatic at first sight. And it was these, apparently less active, areas which were hardest to explain.

It was quite a puzzle for astronomers. All they could see in the region of some vigorous radio emissions was what looked like an ordinary star. It was certainly no different in size from other nearby stars which apparently had roughly the same degree of brightness. So why were the radio emissions coming from just one star, with no evidence of spectacular cosmic collisions to explain them? The mystery deepened when the light from one of these stars was analyzed in the usual way. It was refracted into a spectrum, so that the star's chemical composition could be identified by the Fraunhofer lines, and its speed and direction of travel by the amount these lines were shifted toward the red or blue. At first sight there did not appear to be any of the familiar patterns of Fraunhofer lines shifted toward either end of the spectrum. Everyone was initially baffled by the star: it appeared to emit radio waves for no obvious reason, and to have none of the conventional chemistry which had always been detected in all other stars. But, however exotic and bizarre this made it, no one wanted to call the star by anything other than its conventional, impersonal and rather uninspiring scientific label; the code name it was given like all other stars. In this case it happened to be 3C273.

The first person to realize what was going on was Maarten Schmidt, a

Dutch American, who in 1963 showed that the chemistry of 3C273 was not, after all, so different from that of other stars and galaxies. But, far from bringing 3C273 back into the realm of the expected in astronomy and cosmology, his discovery made it even more astonishing than the idea of alien intelligences in other parts of the universe. Schmidt realized that the conventional Fraunhofer lines were there; but they had been so far red shifted that they were almost out of the red end of the visible light spectrum and into the infrared.

This meant that, whatever 3C273 was, it was hurtling away from us at the unbelievable speed of something like 47,400 kilometers (29,625 miles) per second. In astronomical terms, only light has been seen to travel faster. Which in turn meant that it was now several billions of light years away. And this had extraordinary implications. It meant we were seeing the oldest phenomenon in the universe ever observed by a telescope; something that was so far away that, to have the appearance of a nearby star, it must involve a phenomenal amount of energy emitting intense heat and light radiation. Otherwise, how could its light reach the Earth looking as bright as a star whose light had to travel only a fraction of the distance? So, whatever it was, it was exceptionally old and distant, exceptionally energetic and bright— and, by all calculations, given those characteristics, surprisingly small for such a powerful object. It left astronomers completely mystified. Because 3C273, and several similar objects which were discovered soon after it, all had the superficial appearance of stars, they were described as "quasi-stellar." Which led to them acquiring the now more familiar name of "quasars."

Quasars and black holes

The mystery of the quasars set astronomers and cosmologists thinking. Did theories, such as General Relativity, contain within them some clue as to what quasars were? How did they fit into the whole evolutionary picture of the universe? Was there a whole new aspect of cosmology which would force us to change our model of the universe as radically as Newton changed Ptolemy's; or as radically as, in turn, the Big Bang theory revolutionized Newton's ideas? The discovery of the quasars led to the first high-powered conference to bring together both the best astronomical observers and experimentalists in the world, and the top theorists. It was held in 1963 in Texas, and so, not surprisingly, was called the first Texas conference. Rather more surprising is the fact that subsequent meetings were called the second, third, fourth and onward Texas conferences, despite the fact that they were not held in Texas!

One of the big debates amongst theorists at the time of the first Texas conference concerned the aspect of Einstein's mathematics which appeared to predict circumstances in which all matter would collapse in on itself. (It was these equations which Roger Penrose and then Stephen Hawking were later to develop in their singularity theorems, mentioned in chapter 5.) Some years earlier, the American physicist Robert Oppenheimer had caused quite a stir when he queried the way Einstein had formulated his equations to describe these theoretical collapses. But Oppenheimer and Einstein had never discussed Oppenheimer's ideas, for a variety of reasons. First of all, Oppenheimer had been side-tracked from theoretical physics in order to work on the American Atomic Bomb project at Los Alamos. And even after the Second World War was over, he remained preoccupied by the atomic bomb, with the Cold War keeping the project alive. So, by the time Einstein died at Princeton in 1955, Oppenheimer had not really had a chance to put his case to him.

In many ways, Oppenheimer's objections to Einstein's mathematics were very similar to those raised by Lemaître about Einstein's use of the cosmological constant to counteract the expansionary implications of his own equations. Einstein had clearly foreseen that his mathematics implied the collapse of matter into a single dense point under certain cir-

A typical optical telescope picture shows a galaxy in the foreground, and numerous stars or galaxies much farther away from us. Theoretically at least one of these smaller points of light might be something else; starlike but not a star. A radio telescope image (inset) might expose the truth. This one shows a strong source of radio emission as a bright disc at the top left of the picture, and a huge jet of gas spreading about 1.2 million light years away from it toward the bottom right of the picture.

cumstances, but for some reason refused to believe that this could happen in reality. So Einstein had argued that, at a critical density, when matter had been compressed to its maximum, it would resist the inward pressure of the collapse, thus bringing the collapse to a halt. Oppenheimer had urged that this limitation on the theoretical collapse be put aside, and the implications of an unmediated collapse worked through. It seemed that Oppenheimer was saying that Einstein had once again (as he had done with his cosmological constant) put an unnecessary brake on his equations. But, since Einstein and Oppenheimer had never been able to discuss the question further, one can only speculate as to whether Einstein might once again have owned up to making a great blunder, as he had to Lemaître.

In 1963 the theorists who were going to the Texas conference knew that Oppenheimer would be there, and were naturally intrigued to see if his

---

J. Robert Oppenheimer (1904–1967) ended up being best known for leading the Los Alamos part of the Manhattan project, which built the first atomic bombs. He also made a substantial contribution to theoretical physics.

Senator Joseph McCarthy in full flow, making allegations about people he "condemned" as Communist sympathizers.

ideas would be taken up and developed. But there were additional reasons, besides working out whether Einstein was right, for the particular significance of Oppenheimer's work. The quasars, the basic reason for calling the conference, had been seen to involve exceptionally high levels of energy. And the theoretical mathematics which Oppenheimer had been looking at also involved extraordinary energy. It was natural to contemplate the possibility of a connection between Einstein's equations and the power of the quasars. Perhaps Oppenheimer would be able to expand on his earlier work and somehow point to a possible explanation of the quasars?

Unfortunately Oppenheimer stayed very much out of the limelight at the Texas conference. One possible reason that has been advanced is that he was deeply depressed, disillusioned by the politics he had found himself involved in as a result of the Atomic Bomb project. American sensitivity about communism was at its height in the early 1950s; and one of its most unsavory aspects was Senator McCarthy's hounding of public figures suspected of Communist sympathies. Oppenheimer found himself caught up in all this, labeled a security risk because he opposed the construction of the hydrogen bomb, and is said to have felt that his contribution to the American defense program was consequently totally unappreciated. As a result, he may have been reluctant to champion any other controversial causes, even academic ones.

The bomb squad

There were others at the Texas conference, however, who were less reticent, such as John Wheeler, a well-liked lecturer from Princeton, who had also worked at one stage on the bomb project. Wheeler had a reputation for facing up to daunting challenges and taking unconventional approaches. He persuaded people that sorting out the theoretical mystery of the collapses of matter predicted by Einstein might shed light on the observed mystery of the quasars. A number of young cosmologists at the conference left Texas inspired by Wheeler: people like Dennis Sciama and Roger Penrose, who between them would be such an influence on Stephen Hawking (who was not there). Wheeler's student, Kip Thorne, was also there. (He would later work in this area like Stephen, and make a famous bet with him.) They were all destined to become academic theorists who would make invaluable contributions to the mathematics at a later date. The initial work was to be done by others—many of them, ironically enough, from the same immediate background as Oppenheimer: physicists who had been working on the atomic bomb.

What was urgently needed was some evidence that matter could actually collapse in reality, in the same way suggested by some of Einstein's equations. If the collapses of matter he had predicted could only be shown to be possible in theory, as many people thought, then obviously they could have no real presence in the universe and so they would have no relevance to the nature of the quasars. So the first task was to work through Einstein's calculations and see if they were consistent with the considerable amount that was now known about the actual behavior of matter at very high energy levels. This demanded a great deal of number crunching and an intimate knowledge of high-energy physics, which is why Wheeler and his colleagues turned to the physicists who were engaged on the atomic bomb project.

These scientists possessed two essential qualities. Firstly, they were skilled in the use of the giant computers which the American government had provided to calculate the elaborate mathematics of nuclear explosions. These computers were vital in order to work out how to build a bomb, and only a top priority government project could have afforded them at the time. Furthermore, with the Cold War thawing and work on the bomb largely completed, the computers were less fully occupied than they had been. And secondly, these physicists had already acquired considerable specialist knowledge of the nature of high-energy explosions—the sort of phenomenon that had to be at the heart of both the quasars and the collapses of matter predicted by Einstein. It was an incredibly convenient coincidence that the latest technology and the expertise of its masters should be available to carry

John Wheeler had worked under Oppenheimer at Los Alamos on the bomb project, and went on to further Oppenheimer's work on Einstein's equations.

out the complex and lengthy calculations involved; and so physicists at the defense laboratories increasingly found themselves turning their efforts toward cosmology. Hopes were high that they might solve the mystery of the quasars rather sooner than had at first been thought possible.

Closing in on black holes

The immediate challenge was to see if there were any circumstances in which the collapse of a large star need not end in a supernova explosion. If the star were large enough, could it perhaps involve such powerful gravitational forces that nothing would escape the inward collapse? And would all the matter of the star—even its energy—thus be drawn into a denser and denser single point? If so, then this was what Einstein's equations had pre-

---

It took the needs of the U.S. Defense Department to ensure that the earliest calculating machines were developed into the first computers, with data fed in on punch cards by hand.

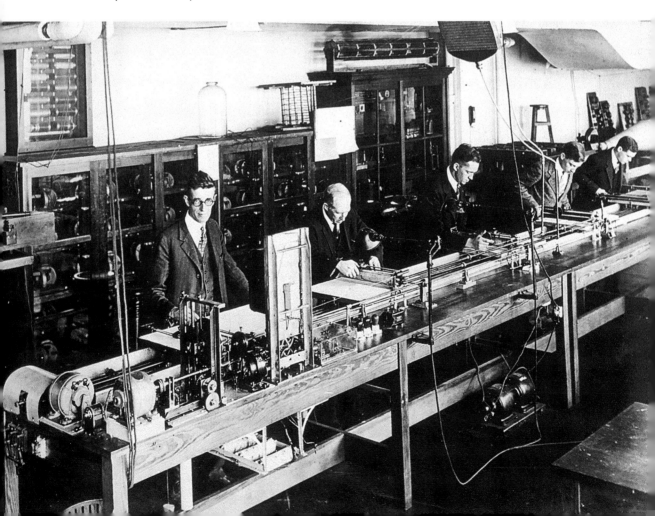

dicted: an awesome, almost unbelievable place in the universe which swallowed all the matter around it, letting nothing—not even light—escape to reveal its presence. John Wheeler chose to describe such a sinister destructive place as a "black hole," and the name has stuck.

The calculations to prove that a black hole could exist were incredibly long and complex, even for the vast computers they were using. In the early sixties these were relatively crude machines, lacking the speed and sophistication of modern microchip electronics. The masses of data required—equations for shock waves, radiation, nuclear reactions and so forth—all had to be laboriously typed onto punch tapes and fed into the computers. But eventually an answer emerged and news of the computers' findings was sent to John Wheeler in Princeton.

According to his students, John Wheeler was an inspiring teacher, not averse to using dramatic methods to encourage his class. When one of his students came up with a revolutionary new thought, he would let off a firecracker. His energy and enthusiasm for his subject was legendary; but he came into his class that morning more agitated than he had ever been before. He had just got the news. "It's all been proved," he burst out. "Black holes can really exist." And he doubtless set off one or two firecrackers to celebrate.

This still did not mean that they actually did exist—only that the known dynamics of nuclear reactions theoretically allowed for a collapse into a black hole to take place in reality. It implied the existence of one or two heavenly bodies massive enough to collapse one day and be pulled continually together by gravity into something that became increasingly dense. There were some stars that seemed large enough; there were certainly whole galaxies which had already been seen in the sky which could theoretically end in such a collapse. But this did not constitute anything like proof that they would actually collapse, or that similar-sized bodies had already collapsed. It would need some form of observation before a black hole could be said to exist in reality. But at least such a strange phenomenon had now been shown to be possible. Moreover, it perhaps underlined an important principle for cosmologists: always allow for the predictions of the mathematics, however improbable they sound.

Einstein had twice found his equations pointing toward difficult possibilities: the expansion of the universe; and the collapse of matter into an infinitely dense point. Despite his conviction that both were impossible and the calculations must therefore be incomplete, evidence had emerged to suggest that he should have had faith in his original mathematics. One of his difficult predictions—the expansion of the universe—had been observed as reality. And now the other had been shown to be consistent with physical laws which had been tried and tested independently of his theories. Einstein had admitted his error in adding the cosmological constant to his equations, when Lemaître had explained his primeval atom idea. No one knows how he would have reacted to the news that black holes could exist in reality; perhaps he would have admitted having made a second unnecessary refinement to his theorems. But in a sense it does not matter. The important point was that cosmologists now realized that in future they would have to resist being skeptical of difficult predictions based on mathematics.

But how could all this help explain the mystery of the quasars? If no one could actually detect a black hole (since it would swallow everything that might give its presence away), it seemed unlikely that a real connection between black holes and quasars could ever be established. The onus was on the theorists to come up with two things. First, they had to indicate what you might expect to see as the result of a black hole, even if you could not

An artist's impression of a black hole, surrounded by stars. The large star is possibly meant to be a binary star partner to the star which collapsed to form the black hole.

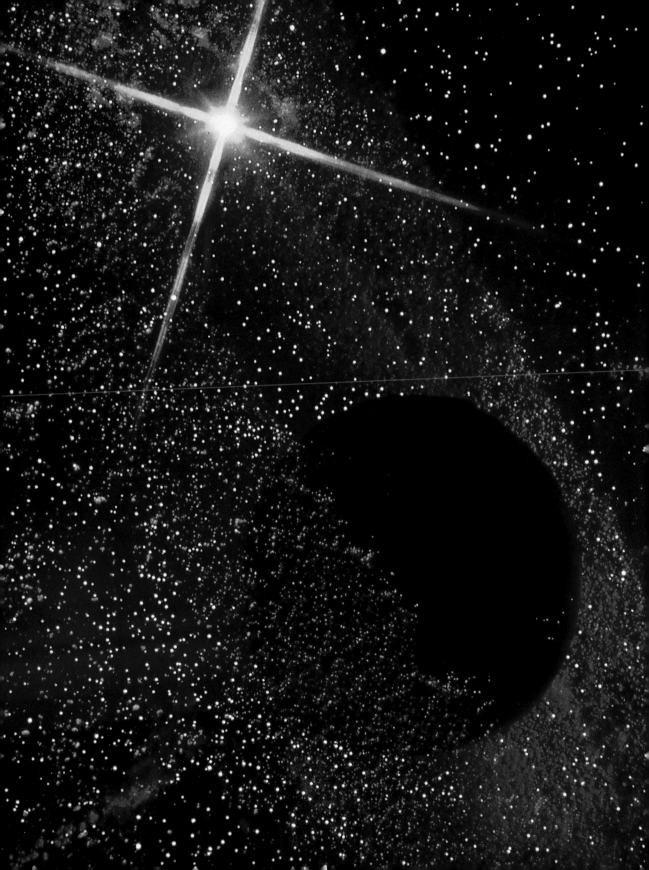

see the black hole itself. This would give observers a way to look for them. And second, they had to propose a relationship between quasars and black holes which could then be looked for with telescopes probing the depths of the universe. If no sensible ideas were forthcoming, the truth about the quasars might easily remain as elusive as signs of extraterrestrial intelligence elsewhere in the vastness of the universe. Radio astronomy had raised some puzzling issues which it was going to prove hard to resolve.

In this artist's impression of a black hole, the so-called accretion disc, glowing hotter and hotter as pieces of matter crash and rub together, can clearly be seen swirling faster and faster around the mouth of the hole; and a huge jet of hot gaseous matter is being shot out into space.

Extraterrestrial Intelligence and the Baffling Quasars

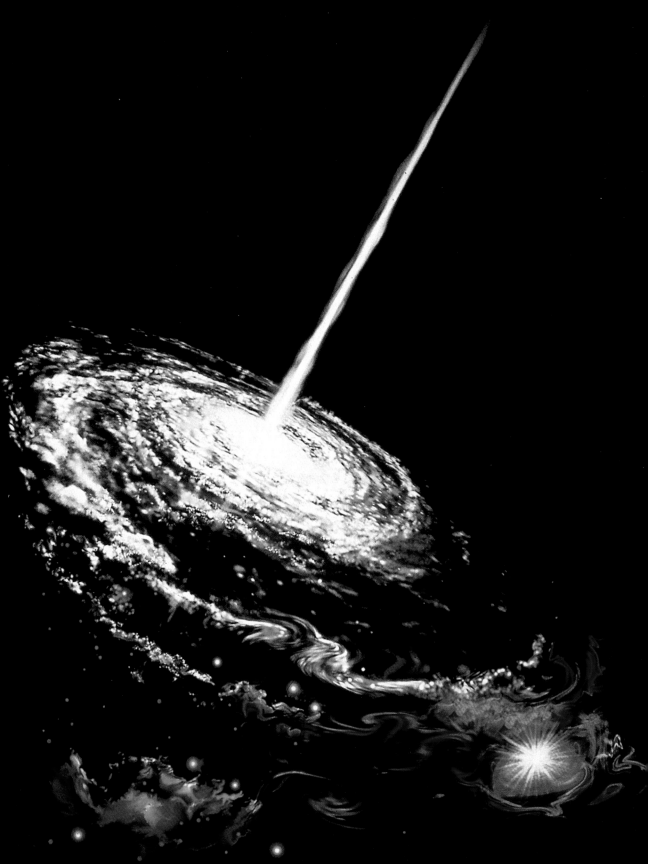

chapter
eleven

～～～

In Search
of Black Holes

WITH ALL THE ATTENTION BLACK HOLES WERE GETTING AFTER THE FIRST
Texas conference, it was hardly surprising that they soon started to appear
in science fiction. A black hole was something with huge energy, sinisterly
devouring everything around it. Lurking unseen, it destroyed stars—even
whole galaxies. It was better than any fantasy scenario that even the most
imaginative writer could conjure up. In any case, it was natural for science

---

A large jet of matter is being emitted from the center of this
artist's impression of a galaxy. This emission would require
enormous energy, and could well imply the presence of a black
hole at the heart of the galaxy. A supernova is shown in the lower
right-hand corner of the picture.

fiction to pick up ideas from science fact. Many of the best science fiction stories gain their plausibility by respecting the findings of real science. And many of the best science fiction writers are themselves scientists. Science fact feeds science fiction; and science fiction is perhaps just as important in supporting science fact. It is certainly possible to argue that in the 1960s science fiction sometimes helped to sustain real science's interest in black holes.

Imagining the impossible

Not long after the first Texas conference, science fact received a jolt which might have led to the abandonment of the quest for black holes altogether. Roger Penrose had been working on the mathematics of the inward collapse of matter under strong gravitational forces, using theorems from topology (the branch of mathematics which studies and explores the properties of

---

Roger Penrose in front of a blackboard explanation of one of his highly visual mathematical theories. This one is vividly called his "twister" theory.

Penrose's ideas are reputed to have inspired the artist M. C. Escher to produce pictures like his *Waterfall* (*right*). At first glance the dynamics of the picture seem to work; but a closer examination reveals something disturbingly wrong with the way the water is flowing.

shapes). Penrose seems to have a special ability to conceive of the way shapes affect each other; indeed, his ideas are said to have inspired the artist M. C. Escher to draw *The Waterfall* and *The Ascending Staircase*. These two famous "puzzle" pictures both reveal seemingly plausible structures which should be impossible in the real world. In one, a continuous fall of water somehow completes a circuit, getting back to where it started—impossible, of course, without the water somehow flowing uphill at some point. And in the other picture, steps complete the four sides of a square whilst seeming always to be going up higher and higher; which they cannot do if they are a continuous circuit, rather like the water in the other picture. It is impossible in the real world to keep going upward from a starting point at a specific level, and come back to the same starting point without going down again to that level; or vice versa.

The relevance of these famous paradoxical pictures is that they show a theoretically impossible event as if it was actually happening. The paradox might be expressed in the form of the dilemma the pictures present for the innocent observer. Is the event real, and so the theory wrong? Or is the theory correct, and the event unreal? Using the same kind of mathematics which inspired the Escher pictures, Roger Penrose produced a similar dilemma for physics. He showed that the inward collapse of matter predicted by Einstein's equations was not only theoretically possible in certain cases; it was an inevitable consequence of Einstein's mathematics that these collapses must happen, and must end up with all the matter involved crushed down into an infinitely dense point. He called this point a "singularity." And, at this point of singularity, it was clear that none of the basic laws of physics would still hold. The mathematics simply did not allow for any other possibility.

This seemed absurd in some ways. He was asking physicists to accept that the very rules for understanding matter and the nature of the universe, which had thrown up this singularity, were now destroying themselves and their validity as a consequence. The theory said there had to be a point where the theory did not work. And if this paradoxical point, a singularity, was the inevitable consequence of a black hole, could a black hole really exist? Or was it only a child of theory, an impossibility that could be given a convincing appearance; an image like an Escher picture, which looked plausible at first but turned out to be an impossibility in the real world?

Then Stephen Hawking showed that Penrose's picture of a collapse to a singularity could be reversed in time. Instead of something huge collapsing to a singularity, a singularity could expand to the largest possible thing—the universe. It would have grown rapidly, expanding from the singularity at the time of the Big Bang. And this, in some ways, made the difficulties for physics even worse. After all, Stephen had shown that the whole universe began from a singularity, and developed according to the established laws of physics from the next fraction of a second onward: until now—some 15 billion years later. And yet, at the point of singularity itself, these same laws of physics could not apply. How could the universe obey laws for its whole existence but not for the first fraction of a second? To this day, no one has come up with an entirely acceptable explanation. The singularity remains an awesome mystery.

Journey to the center of a black hole

Nevertheless, despite all this theoretical work undermining the credibility of black holes, the determination to look for them survived. Perhaps the very realistic images being painted in science fiction helped. The public wanted to know all sorts of things, such as how near to a black hole you can get, and what would happen if you fell into one. Vivid fictional descriptions emerged, based on scientific theory, of what would happen if you fell into a black hole, and what it would be like to be near its very edge. Stephen Hawking's own metaphor, shared by many physicists, is that anyone who fell into a black hole would be stretched out like spaghetti. And in that condition it is doubtful that you would get too worried about the precise nature of the singularity toward which you were hurtling.

With all this speculation, people became much more concerned to find out if such destructive black holes could actually exist. According to the fictional accounts, based on scientific theory, you could not escape once you were inside a black hole. At the very edge of a black hole, however, there would be a life and death struggle going on. It would be rather like drifting toward dangerous rapids in a river. There would be a main current pulling everything toward the rapids; but, as more and more matter got jostled and jammed together in this flow, all sorts of chance happenings would decide the fate of individual bits of matter. Some might even be pushed aside to safety in the calmer side currents; but most would get bundled into a whirling mass, pulled along and crashing into other matter, as it swirled around faster and faster in the ever-accelerating current.

At the edge of a black hole, matter would be drifting toward the cen-

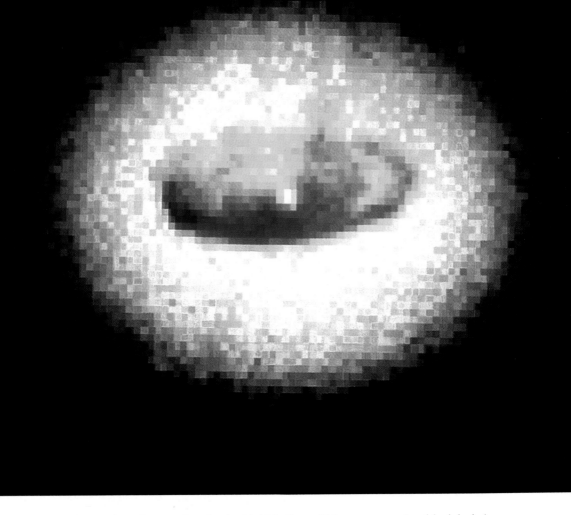

This fascinating picture taken by the Hubble Space Telescope reveals a black hole in the centre of the galaxy NGC 4261. The darker disc is dust and measures a staggering 800 light years across. It is swirling extremely fast around the core, which must be exerting an enormously powerful gravitational pull.

tral pull of its vortex, rather like logs in a stream heading for the rapids. A more accurate, if less dramatic, analogy would be crumbs on the surface of a basin of water emptying down the plug hole. As the crumbs get drawn closer to the water swirling around the plug hole, they begin to circle faster and faster before finally being sucked down. Vast quantities of matter would be orbiting a black hole in the same way, and would be pulled round at such speeds that some of it would inevitably clash violently with other matter. Some of the collisions with other matter would be exactly like those in a particle accelerator. The resulting explosions, as well as the friction and heat caused as matter rubbed across other matter in orbit at a slightly different angle or speed, would all generate colossal energies, radiating outward into space. Perhaps, at some stage, it could even give off enough light to reveal itself as a quasar.

The discovery of black holes inspired science fiction writers and filmmakers to boldly go where science fact has only cautiously gone before.

A dramatic artist's impression of Cygnus X-1, an intense X-ray source within our galaxy believed to be a black hole. Matter is being stripped off Cygnus X-1's companion star HDE226868—shown as a huge white sphere—and then spiraled into an accretion disc before finally being pulled into the black hole.

Believing the impossible

With images like these to fire the imagination, the real nature of the singularity was certainly not as important and exciting for science fiction as the possibility that black holes might soon be discovered. And, whether or not this influenced them, physicists also decided that the theoretical problems posed by the singularity should not hold back the search for black holes. This was partly perhaps because they had begun to believe that, however hard it was to explain, the singularity somehow had to be accepted as part of the real picture of the universe.

Probably the most fundamental argument was that the rest of the evidence supporting Big Bang theory makes it hard to believe that the Big Bang did not happen, even if you put aside for the moment Stephen Hawking's own singularity theorems. And if the Big Bang happened, then, according to the singularity theorems, there had to be a singularity involved. As a result, the approach which increasingly had to be adopted in physics was to accept that the singularity had to be real, even if it seemed counterintuitive and unbelievable. The challenge for physics was to somehow come to terms with the singularity's paradoxical nature. There was even some support for this way of thinking from what is called "quantum mechanics" (the branch of physics which studies the behavior of particles and forces in the subatomic world). Physicists had long accepted that the nature of the very smallest particles of matter raised problems for the laws of physics; perhaps the behavior of a subatomic-sized singularity would in some sense resemble the behavior of subatomic-sized particles and energy? After all, quantum mechanics embraces paradoxes which are as hard to explain as the singularity.

Physicists had already discovered that, in the world of quantum mechanics, small packages of light energy and particles would interact together. The detectors in particle accelerators, you will remember, had revealed energy converting into particles, and vice versa, as the result of particle collisions. One way to understand what is going on is to consider the energy as particles too; and to call the smallest possible package of light energy a photon. And this generally works well when explaining how light behaves in other situations too.

For example, when you shine a torch the beam of light is made up of millions of photons. If you shine the beam toward a barrier with two slits in it, several photons find their way through the slits, producing two new beams of light, each emerging through one of the slits. But if there were only one photon of light to start with, rather than millions, how would it behave? Would it somehow choose to go through one slit rather than the other?

An experiment was set up to find out; and a baffling result was obtained. Although there was only a single photon of light, it was shown to be somehow passing through both slits. This was detected by observing the "interference patterns" produced when one light wave crosses another (when several photons had produced two beams of light, one from each slit, there had been an interference pattern where the waves of light from each slit crossed each other). Now, when a single photon of light was used, an interference pattern was still produced on the far side of the two slits. If the photon had only passed through one slit, it could not have produced interference patterns on its own. It had to have somehow produced light from both slits to create an interference pattern. This meant that the tiny package of light could behave like a particle (in the form of photons, as seen in particle accelerator experiments) and at the same time it could behave like a wave, radiating out in all directions. When the experiment was repeated using electrons instead of photons, the result was even more extraordinary. Once again, the single electron somehow produced an interference pattern, as if it had behaved like a wave and passed through both slits. This "wave-particle duality," as it is called, is now accepted as a factor in quantum mechanics, even though the idea of one thing being two things at once is not something we find easy to come to terms with.

There are other ideas in physics which are exclusive to quantum me-

---

A perfect illustration of the symmetry in an active galaxy. The galaxy is called active when some huge central force—probably a black hole—sends out jets of matter from the galactic center. The two jets are fired out exactly opposite to each other.

chanics. Perhaps the most important for theoretical cosmology is physicist Werner Heisenberg's Uncertainty Principle. This states that it is impossible to pinpoint both the momentum of a particle and its position at the same time. This means that you can never be exactly sure what is going on when you are studying the microscopic world of particles. You simply cannot know how a particle is behaving, and so you have to accept the uncertainty of the situation. And this must include such improbable-sounding events as particles popping in and out of existence. However difficult this may be to accept, it at least seems possible that such strange laws may also apply to something as tiny as a singularity: something immensely dense but smaller than an atom. And, if we refer to quantum theory, some problems in physics may even become less intractable. By applying quantum theory to the singularity at the beginning of the universe, and in particular the Uncertainty Principle and a quantum theory of gravity, Stephen Hawking has suggested, as we will see, one way in which the universe could have got started. This is a long way from trying to decide if black holes actually exist or not—except that anything which helps to explain the paradoxical singularity in turn makes the existence of black holes more plausible.

Betting on black holes

While all these ideas about the singularity were evolving, other theorists were considering what telltale signs would reveal the existence of a black hole. In Russia, Yakov Zeldovitch realized that a large number of stars had been spotted as pairs—binary star systems—in which each star orbited around the other. It seemed obvious that their gravitational effects on each other would explain their movements. This suggested that, if one of the stars had collapsed into a black hole, the remaining star would continue to circle around it. The black hole would still have the mass of the collapsed

Yakov Zeldovitch (1914–1987) was, like his American colleagues Wheeler and Oppenheimer, both a theoretical physicist and someone who got involved in his country's bomb project. His typically Russian biographical notes say he was "involved in some military projects."

star, and so its gravitational influence would still affect the other star. This meant that the star would be seen going round in a circle, apparently all by itself. Spotting such a star would mean choosing a likely candidate and then taking light samples from it on a regular basis. The Doppler shift would change between one observation and the next, as the star circled toward and then away from us. And if a star could be shown to be going round and round in circles from such readings, with no partner star to cause the motion, then a black hole partner would be the most likely explanation.

But where would you begin looking for such an event? With billions of stars to choose from, it would be costly and time-consuming simply to choose one at random and make observations over many months, in the hope that it happened to be twinned with a black hole star. Fortunately Zeldovitch had worked out that the colossal gravitational pull of a black hole on its binary partner would violently strip matter away from the surface of the star and release a huge amount of energy in the form of X rays. While X rays could be produced from another phenomenon, and so they alone would not constitute firm evidence of a black hole nearby, at least they could provide a useful marker of where a black hole might be. So, if X-ray detecting telescopes identified X rays being released somewhere, it would be possible to study that area with an optical telescope to look for a binary system. If there was only a single star, it would mean that it was well worth studying more closely And if it had the right pattern of movement, suggesting an unseen object of a mass consistent with a black hole causing the star to orbit around it, then the combined evidence of the movement and the X rays would make the presence of a black hole by far the most likely explanation. In effect, observers would know that they had discovered a black hole.

Meanwhile, other calculations suggested that there could be even more dramatic signs of a black hole. All the matter hurtling at the outer limit of the black hole would be driven around so fast by the black hole's violent spin that some of it would be thrown off with great force in a straight line extending billions of miles into space: a bit like coffee flying out from the top of a cup as it gets swirled around by someone trying to stir in the milk and sugar without using a teaspoon! So this was yet another effect which a black hole could produce and astronomers could look for. The trouble was that other very dense objects—such as neutron stars which can be created in supernova explosions—could also produce jets of matter. So,

This astonishing radio telescope image shows just how powerful a black hole can be. The whole active region of the black hole is the tiny bright spot, indicating a source of radio waves, in the center of the picture. Then, extending out from it to either side, are two giant jets of matter; huge plumes of gas giving off radio signals as they are blown out about 450,000 light years away from the black hole on each side.

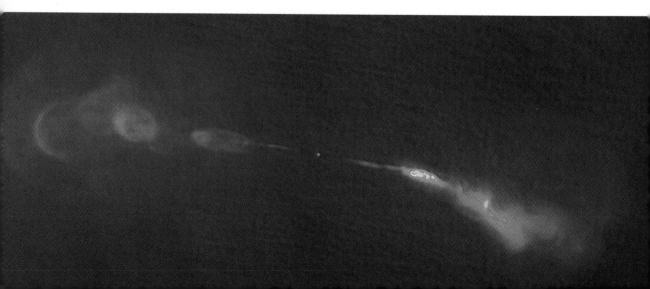

again, they could only be relied on as a marker of a possible black hole; we would need other evidence, like the emission of X rays, and a single star orbiting for no apparent reason, to be sure it was a black hole.

Inevitably, the first interesting observations were incomplete. Huge jets of matter were seen, often emerging from the centre of galaxies; promising X-ray activity was noted in a number of places; but there was nothing conclusive. Everything hung in the balance. Would one of the areas under observation turn out to be a black hole? Or would it turn out that black holes were a fantasy after all, developed from theoretical equations rich in imaginary possibilities but destined to remain theoretical?

Because he was devoting so much of his time to studying black holes, Stephen Hawking made a bet with Kip Thorne, an American working in the same field. (You may remember that Kip Thorne had been at the Texas conference in 1963 as a student of John Wheeler; and Stephen's supervisor, Dennis Sciama, had also been there.) The bet was a bit of intellectual insurance, from Stephen's point of view. He bet that one promising-looking place in the sky would not turn out to be a black hole. Kip Thorne had to take the opposite view; he would win the bet if the place *did* turn out to be a black hole. The stakes, incidentally, were four years' subscription to *Private Eye* for Stephen if he won; and one year's subscription to *Penthouse* for Kip Thorne if he won. Even though Stephen believed that it would turn out to be a black hole, he had bet against it. His reasoning was that, if it did turn out to be a black hole, it would make all his work worthwhile; so he would not mind paying for Kip Thorne to get *Penthouse* for a year. If, on the other hand, it was not a black

Stephen Hawking's betting partner, Kip Thorne, is at the California Institute of Technology. Stephen still visits there regularly, usually at the end of every year.

hole then all his work would have been a waste of time; but at least he would have the consolation of four years of *Private Eye* for nothing!

Gradually the evidence began to mount up. And in 1979 X rays were detected in the region of Cygnus X–1, a star which appeared to be part of a binary star system without a partner. This was the area upon which Stephen and Kip Thorne had based their bet. It was not long before all other possibilities had been eliminated and there was general agreement that a black hole had been discovered. In *A Brief History of Time* Stephen Hawking, writing at the end of the 1980s, says that when the bet was made in 1975, they were 80 percent certain that Cygnus X–1 was the site of a black hole; in 1987 they were 95 percent certain. Within a few years, Stephen finally conceded defeat. He took the note announcing the bet down from its place on a bulletin board, and Kip Thorne started to receive *Penthouse*.

Stranger than fiction

Stephen has, of course, continued to contribute to the theoretical understanding of black holes, exploring their nature by working through the predictions of the complex mathematics involved. His best-known achievement in this field is probably the discovery of Hawking Radiation. Much of Stephen's previous work had defined what happened at the edge of a black hole, the boundary between where light and matter just escape the black hole and where light and matter eventually get drawn into it. One of Stephen's calculations was that the area of activity enclosed by this boundary, known as the "event horizon," would increase as the black hole devoured more and more matter. This was a valuable discovery, since it helped restrict the theoretical options for how a black hole could evolve without violating certain laws of physics concerning energy.

An American research student, Jacob Bekenstein at Princeton, then suggested that the properties of a black hole predicted by Stephen's mathematics made it necessary for the black hole to have a temperature. It was a proposition that Stephen first thought ill-conceived; anything with a temperature radiates heat; and since a black hole, by definition, should not let anything escape from within it, how could it radiate anything? Then, for a completely different reason, Stephen set about calculating how a black hole could emit particles as it rotated. To his surprise, he discovered that the equations he had worked up predicted that a black hole would emit particles independently of its rotation; and that these emissions would be exactly consistent with the temperature Bekenstein's ideas had implied. Stephen Hawking had discovered how a black hole could emit radiation and at the same time obey the laws of physics concerning energy.

While Stephen and others were engaged in the complex theoretical task of defining a black hole, astronomers were beginning to see more examples of black holes; they were being found at the heart of galaxies and with some frequency. Although the quasars were too distant for any detailed observations to reveal their link with black holes, one significant clue was spotted by Susan Wyckoff and Peter Wehinger in 1980. They saw a nebulous region around 3C273, which you will remember was one of the earliest quasars to be identified. The implication was clear enough. The earliest observed nebulae had turned out to be galaxies, once it had become possible to see them more closely. If the quasar was surrounded by nebulous areas, it suggested that it was at the heart of a galaxy too, just like so many black holes seemed to be. The evidence for this was that more and more of the huge jets of matter, which were known to be an indication of a possible black hole, were being observed at the heart of nearby galaxies.

But, if quasars and black holes are in some way linked together, why are we not seeing the light of quasars in any of these nearby galaxies? One very plausible explanation arises from the realization that the quasars are as far away as we can see. In other words, their light is taking far longer to get to us than the light from nearer galaxies. We are seeing them as they were billions of years ago, at the time when galaxies were first forming. We see the nearer galaxies as they were much more recently, in astronomical terms. So they will have evolved significantly beyond the stage when the quasars were at their heart. All this suggests that the high energies of the quasars played an integral part in the early formation of galaxies; but were no longer in evidence later on.

One possible reason is that black holes were a primary source of the gravitational pull which initially gathered matter together into the clumps from which galaxies formed. The nuclear reactions from particles colliding in the swirling, seething streams of matter orbiting the black holes would have radiated the light of the quasars. Gradually, however, the black hole would pull all this colliding matter into itself, and things would quieten down. A lot of matter just outside the black hole would be pulled in opposite directions—the black hole would tug it one way, but stars and gas on the other side of this matter would be pulling it away from the black hole. So the situation would stabilize, with a ring of matter in orbit around the black hole but never falling into it. The collisions in the disc which had produced the light of the quasar would die down leaving only the starlight from all the stars clustered around the center of the galaxy. The light we see in the nearer galaxies is by no means as bright as the light from quasars; it is the light from the central clusters of stars that evolved from this early quasar activity. The central stars in the nearer galaxies are part of an area which is still massively energetic, but far more subdued than the violent energetic clashes which caused the much earlier quasars from which these stars were born.

If this is correct, it means we have gone a long way toward making sense of the mysterious radio signals picked up by telescopes less than fifty years ago. We may not have discovered everything which fired the imaginations of early science fiction writers; but a surprisingly large number of ideas which seemed likely to remain bizarre theories have now been confirmed as scientific fact. There have, so far, been no signs of intelligent life elsewhere in the universe; but the extraordinary black holes, with their mysterious singularities still baffling us, have been identified to almost every expert's satisfaction. And a place in the scheme of things, which now seems eminently plausible, has been found for both them and the quasars.

All these developments mean that, whatever else the mathematics of cosmology predicts, we would do well to consider it seriously, however improbable it may sound to us with our limited experience of reality. Science fiction often still prepares the way for acceptance of what eventually turns out to be science fact. For example, using Einstein's equations, it is perfectly possible to predict changes to the shape of space and time which would affect us in ways we have so far found no way to experience—like time warps.

Most people imagine the universe to be a bit like an ever-inflating balloon, with us somewhere inside it. But perhaps the balloon is hardly inflated at all, and is instead a loose and flexible bag. Perhaps we are inside a universe where time and space can be so bent and flexed that the balloon can be folded back on itself. Eventually two parts of the outer skin could somehow get close enough to each other to be linked by "worm holes"—strange tunnels through space and time through which we might one day be able to move from one end of the universe to another.

The writers of science fiction continue to dangle these ideas in front of us, only a step or two ahead of where science fact is taking us at an increasingly breathless pace. We cannot easily conceive of getting outside our own universe; we have always assumed that the universe, by definition, embraces everything there is. But some scientists as well as science fiction writers nowadays give serious consideration to the possibility of there being other universes—perhaps an infinite number of them. And, if we are to understand how our universe ever got started, it looks as if we will have to open our minds to ideas as difficult and counterintuitive as these.

chapter twelve

~~~~~~

Blowing It Up
Out of All Proportion

THINK OF FRED ASTAIRE AND GINGER ROGERS; OR A SHEPHERD AND HIS sheep dog; or Francis Crick and James Watson, who discovered the structure of DNA. There are countless examples of ideal partnerships like these, where the effectiveness of the sum far exceeds that of the parts. And so it has always been in cosmology. Kepler's theoretical inspiration, for instance,

---

The Hubble Space Telescope reveals what the human eye cannot see; faint galaxies huge distances from Earth. The telescope needs only a thirtieth of the light our eyes have to have to register an image. But even with this kind of technology there is a limit to how far back we can look into the early universe.

needed Tycho Brahe's methodical observations before the true shape of planetary orbits around the sun could be properly understood: a perfect example of how theory and observation have always worked hand in hand to help us unravel the true nature of the universe. Early observation of ships on the horizon showed the Earth to be round; the first theoretical models of the universe had to explain this observed fact. They also had to explain the observed positions of the sun, moon, stars and planets. Ptolemy's theoretical model was the first to do so comprehensively, and was good enough to survive until Galileo's observations proved it was flawed. Newton's theories provided a new model of a static universe; 300 years later Hubble's observations led to the Big Bang dynamic model. And, so far, observations have continued to support this theory: Penzias and Wilson detected the background radiation from the Big Bang; and George Smoot and the COBE satellite discovered the tiny temperature fluctuations within it which would evolve into galaxies.

Little bangs and the limits of observation

Unfortunately for cosmology, however, this powerful partnership between theory and observation is reaching its limit. If we want to trace right back to the very beginnings of the universe, to the moment the Big Bang began,

---

It is hard to envision what purely theoretical models of the way the universe evolved should look like. Even real life events like the drip from a rain-soaked tree into a pool of water below can look very unfamiliar when captured in a split second of time. Inflation theory looks at what happened over such a brief period of time in the early universe, and is often described in terms of bubbles.

there are good reasons why observation can no longer help us. It is remarkable enough that we can still detect the radiation from the Big Bang explosion all around us. At absolute zero, 273° below freezing, heat is undetectable. That is how the point of absolute zero is defined. Anything with a temperature of absolute zero cannot therefore be detected or observed by measuring its temperature. And, after 15 billion years, the heat of the Big Bang was theoretically worked out to be only a few degrees above absolute zero. Yet, we have not only detected and observed this relic of the Big Bang, the background radiation; we have detected tiny temperature variations within it, discovering the origins of the galaxies by observation.

Just as there is a limit—absolute zero—to what we can detect at very cold temperatures, there is also a limit to what we can detect at very high temperatures. Theory tells us that at extremely hot temperatures everything has to be opaque; we will not be able to distinguish shape and structure. It is rather as if everything is hidden within an intense hot fog. So, even if we can find ways to build more and more powerful telescopes, capable of seeing even deeper into space than the quasars, we already know that we will not be able to see right back to the time of the Big Bang. Remember that the farther out into space we look, the longer it takes for the light from what we are observing to reach us. We talk of the distance to far-off objects in terms of light years, the time it takes for the light from them to reach us. This means that, if we could see something from which the light has taken over 15 billion years to reach us, we should be able to see the moment of the Big Bang. But we know this is impossible, however powerful our telescopes become. As we journey backward in time, we will hit this fog of intense heat some 300,000 years before we reach the Big Bang.

In fact the best we can do, in terms of observing the very first moments of the universe, is to see the conditions that should theoretically exist just after the Big Bang. Within the limits of current technology, these conditions can be re-created; but only for fractions of a second in particle accelerators. At the moment of the fastest collisions, the pressures and temperatures briefly created when particles collide match those thought to exist within a second of the Big Bang. But these moments, sometimes described as "little bangs" by physicists, are far too brief to give us more than a few clues as to what might have happened. Even though they mimic conditions which will have existed only a fraction of a second after the Big Bang, there is very little evidence to suggest what would have happened that crucial instant earlier. They give us no idea what could not only create those conditions, but also sustain them and allow something as vast and awesome as the universe to evolve.

What these moments in particle accelerators do confirm is that all that can exist at these extremes of temperature and pressure is pure energy. The particle tracks do not start at the actual moment of the collisions; they arise from a split second of pure energy that the detectors pick up at the point of collision. But this tells us nothing about what must have happened just a fraction of a second earlier. How did the Big Bang get started? What produced the expansion and sustained chain reactions which allowed the universe to grow and grow, so that it is still expanding 15 billion years later? It would obviously be tremendous for cosmology if we could build particle accelerators which created the even more extreme conditions which must have existed just before the "little bangs." Perhaps we could then observe phenomena that would help us answer these questions. But, again, there is a limit to how much we can observe.

The very first particle accelerators, only a meter or so long, could make particles move fast enough for them to break up atoms after accelerating over only this short distance. But, to accelerate particles even faster, you have to let them travel over progressively longer distances in order to achieve every tiny little increase in speed. It takes a journey of about 27 kilometers (more than 16 miles) around the accelerator at CERN to collide particles traveling fast enough to create a "little bang"; calculations of the size

A computer display of a particle collision could be the nearest we will ever get to seeing the Big Bang. Some very violent collisions in particle accelerators will produce, for a very brief moment, the kind of pressure and heat which must have existed within a second of the Big Bang.

of accelerator needed to mimic conditions that fraction of a second earlier, when the Big Bang began, vary quite dramatically. Some physicists argue that something the size of the solar system would do the job. Others claim it would need to be the size of the whole universe! It matters not who is right—either size is totally impractical. Again, it seems that we have come up against the limits of observation.

Building pictures with bricks

It appears, therefore, that the ultimate mysteries of the universe are going to have to be unraveled by theory alone. The obvious danger is that the loudest voices and the most eloquent advocates may prevail, rather than experimentally proved principles of science. At first glance it is easy to assume that, within theory alone, there can be no proof that one idea is right rather than another. But, in fact, there are some aspects of developing a theory which carry within them seeds of the required proof.

Imagine theorists as being rather like children playing with those bricks which have parts of a picture on each side. Since each brick has six faces, they can be arranged in a block to reveal one of six possible pictures. It is natural to concentrate on building one meaningful picture at a time. So, using trial and error, the bricks are placed next to each other like jigsaw pieces, rearranged, and then rearranged again, until a meaningful picture starts to emerge. A good cosmological theory can be pieced together in the same way. The question then is: how can we prove that this picture is the real picture and not simply a convincing-looking possibility?

With the bricks, it is usually possible to turn over the complete set, carefully ensuring that the picture that has been built up is preserved. It then becomes the unseen face, on the bottom of the block of bricks. The face that was originally hidden is now exposed as the top surface. If what you see is another completely clear, meaningful picture, quite different from the one you originally pieced together, then you can be sure that the first picture you built was right. In the same way, building a promising cosmological theory, and finding that it allows another theory to fit perfectly into place, offers some kind of proof that the original theory has to be right.

A good example is Stephen Hawking's discovery of Hawking Radiation. You will remember that he was trying to build an elegant set of equations, a mathematical theory to show how the rotation of a black hole would produce an emission of particles. He wanted to improve on the theo-

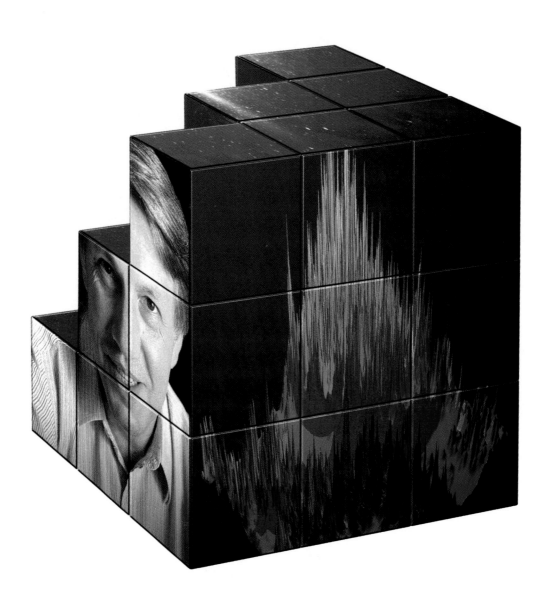

When the pictures on one side of the bricks are arranged to make sense, another picture will automatically appear on the opposite face of the bricks. A kind of proof that the first picture is right. Our bricks show Andrei Linde and a computer picture of the way his third theory of inflation might work.

rems which some Russian cosmologists had worked out. Stephen found that his equations offered a slightly different picture; they showed that a black hole would emit radiation irrespective of its rotation. So, was this picture—the equivalent of the picture on one face of the block of bricks—right or not? When Stephen looked at the consequences of this picture (the equivalent of turning the block of bricks over), he saw that it predicted a temperature for the black hole which exactly matched the temperature needed to make sense of Jacob Bekenstein's theory. This was the equivalent of finding that a second meaningful picture had been created on the unseen face of the block of bricks.

Supported by this kind of proof of their accuracy, one or two purely theoretical ideas have already won widespread acceptance. They go some way toward advancing our understanding of the unobservable fraction of a second at the very beginning of the universe. However, other ideas have been developed with little other theoretical support. So we will need to pick our way carefully between those theories which are broadly accepted by cosmologists today, and those which are only the unproven ideas of individual advocates.

A tiny vacuum

One important idea which most cosmologists find acceptable is something called inflation theory, which its origins at the end of the 1970s. The Cold War was starting to thaw, but in Russia the publication of ideas was still very much under central government control. This was particularly true when it came to announcing scientific discoveries. In Moscow a bright young physicist, Andrei Linde, was working on theoretical explanations of what could have caused the rapid expansion of the early universe which must have happened immediately the Big Bang occurred. There had to be something, otherwise there was no way the universe could end up at its present size, with the rate of expansion we observe today.

One day he had an inspired thought. He remembered that a few years earlier it had been argued that, in principle, energy could spontaneously evolve from nothing at all, if quantum theory was applied to the laws governing the nature of a vacuum. This meant allowing for the Uncertainty Principle to apply to a tiny vacuum, just as it did to subatomic particles. In the same way that quantum mechanics allows particles somehow to pop in and out of existence, the theory would allow the tiniest of vacuums to do so. Since their experience with black holes had shown scientists that you have to respect the possible consequences of the mathematics of good theories, however improbable they seem, it seemed worth giving serious consideration to this idea.

The problem was that, even if you could hypothetically allow a tiny vacuum to pop into existence, containing within it a tiny amount of energy, there was no obvious way in which this could be transformed into all the matter in the universe. This is where Linde's inspiration came in. He remembered that some work had been done on the behavior of energy in a vacuum which suggested that it would spontaneously expand. What would happen, Linde wondered, if the tiny vacuum of energy somehow popped into existence, expanded rapidly and failed to pop out of existence again? Would the expansion of the energy continue and be sustained long enough to allow for the emergence of a universe full of matter anything like the one we observe today?

Excited at the possibility, Linde worked through the mathematics, and saw that a universe could theoretically emerge from a rapid expansion of this sort. But there was a problem—it would not be a universe with all the characteristics of our own known universe. So, was the idea worth taking seriously? Linde was convinced that it was, and he determined to refine the theorems to see if he could make them produce a more realistic universe. He did not, however, consider publishing his ideas at this stage; in the climate

After beginning his academic career in Russia Andrei Linde, one of the founders of inflation theory, now teaches at Stanford University.

prevailing in Russian science at the time, he realized there was no way the central committee would approve publication. They would point out that all he could offer was an interesting-sounding theory which could not be proved, and which did not even produce a recognizable universe as a consequence. Any paper he wrote would be rejected out of hand.

Meanwhile, a physicist in the United States was working on the same theoretical possibilities. He, too, had a moment of inspiration. Alan Guth had worked through the initial mathematics the same night that he had his sudden insight, and rushed into work the next day to see his colleagues. He was clutching the book in which he had worked through the equations, on which he had scrawled a two-word title: "Spectacular Idea." As he explained his theory of inflation, his colleagues could see that it was an important step forward. He too was explaining how an early inflation of energy in a vacuum could get the universe started; and he too had realized that the kind of universe it produced would not match the one we see around us. But, unlike the Russians, the Americans had a tradition of scientific publication when there was a promising mechanism, even if there were problems with its outcome. Guth had no hesitation in publishing in one of the American scientific journals, and he was duly credited with discovering inflation theory.

Irrespective of who was actually first to have the basic insight, it is

Alan Guth has the unusual distinction of receiving plaudits and public recognition for an idea he admitted he could not really make work! The principle of inflation he described is still recognized as an important way to help explain the early universe; it is assumed we will one day know precisely how it fits in.

certain that both Linde and Guth came up with inflation theory totally independently. Linde now gives Guth credit for having the confidence to publish the theory first, when both of them acknowledge that it could not actually deliver the basis of the known universe. But undoubtedly the way science is done in both countries also influenced the decision to publish.

Blowing bubbles

Many cosmologists now believe that inflation theory does account for one aspect of the early universe; but several think that other, totally different theories will be needed to explain its initial failure to make a convincing model of our universe. Linde, however, was convinced that inflation theory could be refined to do away with this problem. And one night he suddenly thought of a possible solution. He was talking on the telephone at the time. So as not to disturb his wife, who had already gone to bed, he had taken the phone into the bathroom. Whatever the conversation was about, Linde abruptly stopped paying any attention as it suddenly struck him that there could be another approach to inflation. Linde dropped the phone, leaving his friend stranded at the other end of the line, and dashed into the bedroom to wake his wife. Excitedly he told her: "I think I know how the universe got started."

The key thing which had dawned on Linde was that he had previously assumed that all the energy in the vacuum would have inflated into the primeval universe. But what if it did not happen like that? Suppose all the energy in the vacuum was released, rather like the gas in a carbonated drink when the bottle is opened. In the drink, many bubbles of gas—not just one big bubble—rush to the top of the bottle when the cap is unscrewed and the pressure is released. Why not imagine that the energy in the vacuum breaks up into billions of tiny bubbles in a similar way; and that only one of these bubbles expands to become our universe? When Linde worked out the equations, he found he could now produce a universe from

Inflation theory argues that energy in a vacuum will expand and be released, rather like bubbles of gas in a carbonated drink. The problem for theorists like Linde and Guth has been to work out how the universe we see around us would evolve from that point.

inflation which was compatible with the one we live in. It seemed that his boast to his wife had been justified.

But there was an obvious question still to be answered. If only one tiny bubble of energy in the vacuum inflated to produce the universe, what happened to all the others? Each bubble, after all, would have the same potential to inflate into a universe as the one which had evolved into our known universe. Logically, there were two choices: either each bubble becomes a universe, outside our own, which we can never see or detect; or some extra factor comes into play which eliminates all the other bubbles so that only our universe survives. And, since there was no natural way for this factor to arise from the equations which predicted and explained the inflation, it seemed that a new law of physics would have to be found.

Yet again the Russian idea of what constituted good science delayed Linde in publishing his multiuniverse version of inflation. Although he eventually did get his theory out first, two other Americans working on inflation theory published a paper explaining this idea, and its shortcomings, far more easily than Andrei Linde could. The Russian began to despair of ever being allowed by his own authorities to practice theoretical physics as openly and easily as his colleagues in the West. But, by now, cosmologists worldwide were aware of Linde's ideas, and he was duly credited with having come up with this second version of inflation theory. It was not, however, all that popular with other cosmologists, Stephen Hawking among them.

In 1981 Stephen attended an international conference in Moscow, where the interpreters had difficulty in understanding him. His voice was quite noticeably deteriorating because of his illness, and the technicalities he wanted to discuss involved a specialist vocabulary which only another cosmologist could be expected to understand easily. So Andrei Linde agreed to act as interpreter while Stephen gave his paper to the conference. Linde recalls being somewhat embarrassed as Stephen's line of argument unfolded. He was putting Linde's multiuniverse ideas into perspective by pointing out their shortcomings. And, as a result, Linde was having to spell out the weaknesses in his own theory to his Russian colleagues!

Apparently Hawking and Linde then went off together and worked through the equations in Stephen's hotel room. There was no real dispute between them in the end, as Linde was already aware of his theory's shortcomings. One problem with having all the bubbles of energy create separate universes is that, mathematically, there has to be an infinite number of possible universes; this means that, somewhere, every conceivable scenario is taking place. Instead of narrowing down the options and explaining how our universe turned out to be the way it is, the theory simply says our universe is the way it is because all possibilities have to exist, including our universe. We already know we are here in this universe from our current observations; the multiuniverse theory cannot help us understand any more than that. It is a theory that allows for all things to be possible, and so predicts nothing. Anything that might happen is allowed by the theory and so the theory has no significance.

The alternative interpretation of the theory, allowing only our universe to survive, is equally unsatisfactory. When no established laws of physics appear to allow for this possibility, simply inventing one to do so is just as destructive of the theory's credibility. It is the equivalent of saying that all possibilities exist but only our universe survives because some law makes it survive. Again, the theory is effectively exposed as predicting nothing.

Pulling a rabbit out of a hat

Andrei Linde recognized all this, and at the same time was finding it increasingly difficult to cope with the bureaucracy of Russian science. He became deeply depressed, and found it difficult to maintain his enthusiasm for his world. His imaginative commitment to making sense of inflation theory was beginning to wilt when he received an instruction from on high. He was to attend an international conference on physics as a representative of the U.S.S.R.; and he was told a few days beforehand that he was expected to deliver a paper on something significant, to demonstrate the quality of Russian science. This concentrated his mind, and helped him, as he puts it, to pull a rabbit out of the hat. But it also made him finally resolve to quit the Soviet Union and continue his work in the U.S.A.

The rabbit he produced has been refined into a third theory of inflation, which Linde proudly explains with some colorful graphics on a computer at his new home in Stanford, where he and his wife both now teach physics at the university. It is a revolutionary approach which is supported by few other cosmologists, but which he bravely claims removes all the problems of the Big Bang and identifying a moment of creation. It depends on the idea of fields in physics.

Most people realize that the properties of magnetism are revealed in a magnetic field; and the movement of an electrically charged particle will depend on its position in such a field. Linde suggests that in another kind of field, called a scalar field, there could be tiny bubbles of energy being produced and inflating into universes all the time. There would be groups of such universes with similar properties, which would all be interlinked, rather like the bubbles on the surface of boiling water. Within each group, depending on how the scalar field had thrown them up, some universes would develop faster or slower than others; and our universe would be one universe in one group which happens to have developed at exactly the pace we observe.

Linde believes he has discovered a pattern for the continuous evolution of an eternal network of universes, each one developing naturally from a scalar field. There will be a Big Bang as a part of the way each universe evolves, but it will not be fundamental, avoiding all the problems the singularity seemed to pose. However, he has yet to convince many other cosmologists of every aspect of his idea. He may be out on his own with his scalar field theory; but the basic idea of inflation from which it all started is largely accepted as a likely factor in explaining the earliest moments of the universe. The problem is finding a way to prove it right or wrong. While nothing has so far emerged to confirm it, nothing has yet proved it wrong either. Inflation theory still has to offer, as it were, a different convincing picture on the opposite face of the block of bricks. On the other hand, no one as yet has offered a better explanation of how the early expansion of the universe arose.

chapter
thirteen

~~~~~~~~~~

Everything Tied Up
with String

Survival of the fittest

WHEN ANDREI LINDE FINALLY LEFT RUSSIA FOR THE UNITED STATES, HE was still pondering the problems of his second version of inflation theory, one of which was that he needed a new law of physics to get rid of the un-

---

Satellites have enabled scientists to look out into space without the interference of the Earth's atmosphere corrupting the data they collect. The Hubble Space Telescope and COBE are good examples. With a lot of ingenuity new satellite experiments may yet be devised to shed light on the origins of the universe.

wanted extra universes which the model predicted. Linde ended up at Stanford University on the West Coast of America. On the eastern side of the continent, in Pennsylvania, another cosmologist was thinking about new natural laws which might apply to cosmology. But his starting point was not the quantum world in which Linde had found tiny packets of energy inflating in a vacuum. Lee Smolin was working on the implications of Einstein's mathematics, thinking about black holes and singularities, and how Stephen Hawking had shown that the universe must evolve from a singularity.

He thought about the number of singularities which must exist in the universe as a result of all the black holes which were being discovered. Why did all of them, or at least some of them, not develop into new universes? He could not find a law of physics to explain this neatly, and so he did a bit of lateral thinking. There were situations in nature where every possible starting point did not give rise to whatever might grow from it. In biology there are countless examples. Fish lay millions of eggs, to allow for the fact that only comparatively few will be fertilized, and many of the resulting fry will be eaten or fail to reach maturity for some other reason. In humans and most mammals, it takes millions of sperm to ensure that just one reaches the end of a hazardous journey and fertilizes an egg from which a baby will grow. It is all part of a biological law commonly known as the survival of the fittest— the central phenomenon in the dynamics of the evolutionary process described by Charles

Lee Smolin wonders whether biology can help physics explain why we are in the kind of universe we are in.

Darwin which has given us such a rich, complex pattern of life on Earth.

Lee Smolin wondered whether some similar law might apply to the evolution of the universe. Perhaps in the whole history of the universe a number of singularities could be produced, but only one would be destined to survive, like the lone sperm which fertilizes an egg. Only the fittest singularity would end up producing the universe.

A theory of everything

Even though Lee Smolin's philosophy seems to offer an interesting solution to Linde's multiuniverse problem, it has not won much support

---

Can nature reveal anything that would help make sense of cosmological problems? Huge numbers of fish eggs are needed to ensure that enough survive and mature to keep the species going. In the same way, could our universe be the sole survivor of a whole batch of embryonic universes?

Albert Einstein contributed perhaps more key ideas to cosmology than any other scientist—and yet he was unable to unite all physics with a Theory of Everything.

from cosmologists. As a theory, it does not easily lend itself to being embraced by the equations and proofs of physics. Most cosmologists naturally want to find an all-embracing description of the universe within the laws of physics. They have come so close to achieving this ambition that it must seem unnecessary to most of them to call in the laws of a completely different scientific discipline. But, equally, physicists realize that there is a central problem which has to be surmounted before their science can give a complete explanation of how everything works. And that is the mismatch between the physics of the very large (relativity) and the physics of the very small (quantum mechanics). It is a bit like drilling a huge tunnel. If you start work from opposite sides of the mountain studiously progressing according to well-drawn plans, you know you have a problem when both carefully controlled tunnels just will not meet in the middle.

Einstein's General Theory of Relativity, at one end of the tunnel as it were, does a marvelous job of explaining the large-scale dynamics of the universe. Its equations for gravity work so well that the observed orbits of all the planets in the solar system can be precisely predicted, despite the huge distances involved. If we can discover the right quantities of dark matter, with all the right properties, then it will also perfectly explain the movements of whole galaxies. At the other end of the tunnel, quantum mechanics does an equally impressive job of describing the behavior of subatomic particles. Its theoretical rules, such as Heisenberg's Uncertainty Principle and the idea of wave-particle duality, are precisely confirmed by what we observe in particle accelerators. But, when physicists try to marry these two great principles of physics at the point where they should meet, they do not seem to mesh. The large physics of the universe (governed by gravity, and springing from a tiny singularity) needs to embrace the small physics of quantum mechanics in order to explain how that singularity can arise and give birth to the Big Bang, and the gravitational effects which shape the universe from it.

The search for this relationship between both sides of physics has been called the search for a "theory of everything." The assumption is that, once it has been discovered, it will enable physics to explain how everything works in the universe. It will embrace all the known forces at work, and how they are involved in both the workings of the atom and the dynamics of the cosmos. Einstein realized the importance of discovering the equations involved, and was convinced that they might end up being simplified to something as brief and immensely significant as his earlier equation, $E = mc^2$. He spent his last years at Princeton University, working almost exclusively on finding a theory of everything. On the day he died, his desk was still littered with papers on which were written scores of equations. But no one to this day has seen any sign among all these calculations that he was close to discovering what he was looking for.

Strings of quarks

It was a few years after Einstein's death before a promising possibility emerged. In the early 1960s, particle physicists had started to work out that there had to be a more fundamental set of subatomic particles than the ones regularly seen in particle accelerators. Eventually they predicted a set of six such particles, which they called quarks. The idea was that they all had different properties, and they could be arranged in different groups of three to create the basic characteristics of particles at the next level of matter. For a while they were only the prediction of a well-founded theory; but once there was actual observational evidence supporting this theory, physicists were puzzled at first. For some reason it seemed impossible to see individual quarks in isolation; they were somehow always linked together.

The idea grew that the bigger particles, which were made up of quarks, were a bit like little pieces of string, with a quark at each end of the string. Sometimes the strings were left as a single length; sometimes they would have the ends joined to form a tiny loop. Either way, the quarks were inseparable because they were parts of the string. And, depending on the nature of the three quarks which made up each of the little lengths and loops of string, the string would vibrate in a distinctive way, determining the characteristic behavior of the particle they combined to create. For this reason the quarks were given very unscientific-sounding names, to describe in broad terms how they contributed to the dynamics of a particle: names like "top" and "bottom," "up" and "down," "strange" and "charm." Instead of particles being considered as single points, they were now seen as vibrating strings whose vibrations contributed aspects of the forces detected in atoms.

Strange as this string theory might sound, it considerably simplified the calculations involved in working out the role of particles in building bigger structures. All of this might have been of no consequence to cosmologists were it not for the kind of mathematics which best embraced this new way of looking at subatomic particles. This was topology, the very same branch of mathematics from which Stephen Hawking and Roger Penrose had built their theorems to explain the singularity at the bottom of black holes and at the beginning of the universe. This had, in turn, depended on Einstein's equations describing gravity. Topology thus embraced both the mathematics of relativity and gravity, and the mathematics of subatomic particles as expressed in string theory. Did this mean that string theory and topology in some way contained the elusive theory of everything, which would unite physics and perhaps ultimately explain the start of the universe?

Everything Tied Up with String

Unraveling the inner workings of the atom, it becomes increasingly difficult to see what is going on. Each consists of a nucleus surrounded by electrons (*above*). The nucleus is made up of protons and electrons, each of which is made up of groups of three quarks (*center*). Any set of three quarks—these (*right*) make up a proton—would have strings at their heart. To get an idea of comparative size, imagine that an atom is the size of the solar system. By comparison, a string is the size of an atom.

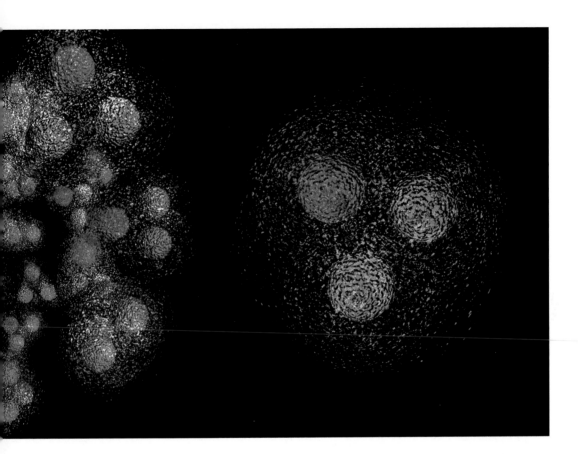

The eleventh dimension

One story has it that, quite by accident, while passing the time on a ferry trip in Greece, a physicist was thumbing through a mathematical encyclopedia. First he spotted the formula for gravity among hundreds of equations in the branch of topology which includes string theory. Then he saw equations describing the electromagnetic force which is central to subatomic physics. However the connection was made, physicists in search of a theory of everything quickly took on the challenge of trying to find a way through the numerous complex equations involved in string theory. At first there was a lot of excitement, and high hopes that they would eventually work out a formula to explain the first moments of the universe. But, by the early 1980s, no really promising leads had been discovered, and the full complexity of string theory mathematics was beginning to dawn on the physicists working with it.

Essentially, since the vibration of strings involved movements through space and time, the mathematics of strings had to be considered in at least four dimensions. First, there are the three dimensions of space: think of an object such as a box which has height, width and depth. Then there is the extra dimension of time. This fourth dimension is not too difficult to grasp if you imagine the box being carried from one room to another. It has not changed its height, width or depth at all, but it has changed its overall position in the time it takes to pass from one room to the next.

Unfortunately, the mathematics of strings is not easily confined to these four dimensions. Imagine making changes to the box. It is easy enough to imagine varying the height, width or depth, or the way its position changes through time. But can you imagine a totally separate way for the box to change in addition to these four? Most of us find this extremely difficult. Yet the mathematics of string theory predicts a fifth and many other types of changes. The easiest way to talk about these different facets of behavior is to call them the fifth, sixth and seventh dimensions, and so on. Difficult as these dynamics are to imagine, we have to work with them in order to sort out the equations of string theory. Currently, the majority of top string theory experts accept that there are at least eleven dimensions involved.

This, in itself, makes the calculations complex enough. Even more discouragingly, the first Herculean efforts to find a way through the maze of thousands and thousands of equations seemed to lead nowhere. It began to seem to some physicists that they had lost themselves in a vast, directionless morass of numbers. Rather like the theory of an infinite number of universes, it would predict all possible outcomes and thus fail to offer a meaningful approach to any problem. After the initial excitement, disillusionment inevitably set in. Several cosmologists were amongst the physicists who lost faith in the power of string theory to help unravel the last secrets of the universe. String theory might have been buried and forgotten without the persistence of a few determined supporters who kept searching for a way forward.

M theory

One of those who did not give up on string theory was Ed Witten, a professor at the Institute of Advanced Studies in Princeton. He is known, somewhat irreverently, by other physicists as "the Pope," a clear recognition of his abilities and the position he holds in the field. If anyone could find their way through string theory, Witten was the man to do it, in the eyes of his peers. And in the 1990s he has revived interest in the possibility that a meaningful theory of everything will emerge from the complex web of string theory. He has argued that a lot of the equations involved are a kind of mirror image of other equations. He calls these pairs of equations "dualities" and has set about trying to identify them. They typi-

Ed Witten is burdened with the hopes of others that he will find a way through string theory to reveal a Theory of Everything within it.

It is hoped that the next satellite to examine the background radiation from the Big Bang will be able to see an enormous amount of detail. If what it sees is either compatible or incompatible with patterns predicted in computer models, then we could get some idea of the plausibility of some theories, like inflation and super string theory.

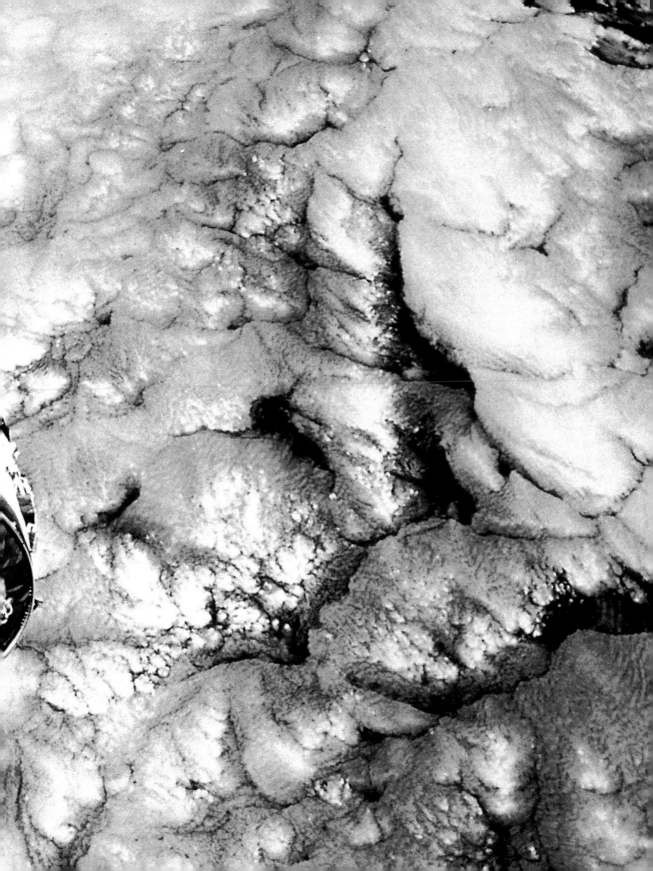

cally turn up in different dimensions; usually with a contrasting role in each. For instance, a strong influence in one dimension would have as its duality partner a weak influence in another.

Witten started to see what would happen if he, in effect, started pulling together as many dualities as he could find. Perhaps if they could all be tidied together, it would help expose a common core, a "central tree trunk" of string theory, as it were. And he believes that a much more meaningful picture is emerging. He calls his way of working with string theory "M theory," which is itself a refinement of a refinement of string theory called "super string theory." He is confident that eventually, when all the mathematics have been carefully simplified, an equation of a manageable length will emerge which does indeed unite both sides of physics: a theory of everything which will explain the dynamics of the beginning of the universe.

But, if and when it does, how will we know for sure that it is in some sense the final answer to an age-old question? We can only hope that there will be "a picture on the other side of the bricks," another theory that fits exactly as a consequence of the equation which emerges from M theory. But there is no guarantee that there will be, and it is quite conceivable that any theorem that emerges will receive only a cautious, provisional acceptance while it is repeatedly probed and tested. The strength of science, after all, is that nothing is considered true without experimental confirmation; and finding an experiment to confirm a theory of everything might just turn out to be a little difficult.

The Planck Explorer project

Incredibly, despite the limits to what we can observe of the universe, there is an experiment under way which might at least help eliminate any wrong ideas. Another satellite is being built to repeat the experiment carried out by

COBE, but with even more sensitive detectors than before. Of course it will not be able to look any further back in time, or get any nearer to the Big Bang itself. But it should be able to study the background radiation which COBE detected, and perhaps pick out even smaller temperature variations. The name of the projected satellite, Planck Explorer, takes its name from an eminent German scientist—no acronym this time!

The satellite is being built by ESA, the European Space Agency; and it will produce a computerized map just as COBE did. That map will be compared with a number of computer maps which have already been built up as theoretical models of the universe, each assuming that one particular theory of the early universe is correct. To draw up these maps, theorists carefully incorporate the mathematics of a particular theory into a computer model of how the known universe would have to have started, and how it would have evolved, if the theory is correct. And in each case a distinctively different map of the background radiation is predicted. One theory will need a number of large hot areas, say, if it is going to produce a universe like ours; while another will only work if there are large clusters of tiny hot spots.

Take Andrei Linde's version of inflation theory which produces an infinite number of universes of which ours is one. The computer modelers will build a model of the background radiation predicted by this theory. Then, once Planck Explorer has given them the real map they can see whether this observed map is compatible with the theoretical map of what would be seen in a universe created from a multiuniverse inflation. At the very least, it may be possible to point out crucial differences in the two maps. The theory may require patterns in the background radiation which are simply not part of the reality observed by the new satellite. In this way, Planck Explorer may help eliminate a number of theories of the early universe, perhaps leaving one plausible one. Computer models which have already been built have shown that the map of the background radiation will look quite different in four different situations. For instance, variations in

Professor Neil Turok, like Hawking, works at Cambridge University. Closely involved with the Planck Explorer project, he wears protective clothes to visit the clean room where the satellite is to be built.

the background radiation map in a universe developed from string theory are clearly different from the map we would find in a universe developed from the multiuniverse theory.

But, whatever the results of the Planck Explorer project, the best we can hope for is proof that some early universe theories cannot be viable. Just because a particular theory is seen to be compatible with the observations of the satellite does not mean that it is the only theory that could be compatible. As we get closer to understanding how the universe works, proving any particular theory correct gets progressively more difficult.

This leaves cosmologists both united and divided. They are united in agreeing, for instance, that one theory is promising, and looks as though it might make a useful contribution. They will perhaps agree that another theory is a revealing way to approach a particular problem. But they are usually careful not to reject any intelligent idea out of hand. Most of them will therefore acknowledge the significance of both inflation theory and string theory in the search for an explanation of the early universe. And many will agree that we are close to finding that explanation. But there the agreement tends to stop. A number of cosmologists' individual ideas combine elements of other theories, but introduce refinements of their own which have so far failed to win support from their peers. And Stephen Hawking is no exception.

chapter fourteen

~~~~~~~

Stephen Hawking's Universe

WHEN GEORGE SMOOT RELEASED THE PINK AND BLUE PICTURE HIS computer had drawn to reveal ripples in the cosmos, it made the front pages of newspapers all around the world. It is difficult to recall an experiment in cosmology achieving such prominence in the media; and the reason it did was partly because of a quote from Stephen Hawking which supported the picture in most newspapers. He did not mince his words. It

---

Stephen Hawking's idea for a "no boundary" universe suggests that the universe does not need a moment of creation, but is completely self-contained. But it is still a dynamic universe in which galaxies like the unusual spiral, NGC 3718, can evolve.

could, he claimed, be "the greatest discovery of the century—if not of all time."

It was April 1992—four years since Stephen had published *A Brief History of Time*. Following the unprecedented success of the book, Stephen was now world-famous, and public interest in cosmology was at its height. So he had no need to make dramatic press statements to bring attention to himself or his book. He was simply genuinely delighted with the success of COBE. It certainly gave great support to Big Bang theory, which was about as far as most newspapers took their front-page stories. But it did something else which was also of great importance for Stephen. It showed that tiny fluctuations existed in the early universe to create the temperature differences COBE had detected in the background radiation. These were the fluctuations which eventually gave birth to the galaxies and voids in the universe we see today; fluctuations which had to be there if Stephen's own proposed explanation of the earliest moments of the universe was right.

Quantum gravity and imaginary time

Stephen has always worked in cosmology with the physics of the very large. His singularity theorems were a direct result of Einstein's mathematics; and, while they gave considerable support to Big Bang theory, they also had a disturbing consequence. As we have seen, the laws of physics which predicted the singularity could not hold at the point of the singularity. This made it very difficult to understand how any basic principles of physics could explain the first moments of the universe. Stephen worked on this problem and came up with a strikingly bold solution. It even seemed at first sight to be a flat contradiction of his earlier work. Having shown that the universe had to have started from a singularity, his solution to the problem it posed appeared to find a way for the effects of the singularity never to be part of the way the universe evolves!

Stephen's proposal for the evolution of the universe essentially begins with the recognition of the primary importance of gravity. The way gravity affects everything which moves through space and time is a clear central prediction of Einstein's General Theory of Relativity. But what happens to gravity in the quantum world, in the physics of the very small and the very dense? The most reasonable assumption seems to be that the same laws of quantum mechanics which apply to subatomic particles ought to apply to anything else of subatomic size. And, since the very early universe had to contain immensely dense matter and forces, all very small, then gravity and everything about the universe would have to be governed by quantum laws at that stage. There needs to be, therefore, a quantum theory of gravity which links the way we understand gravity in General Relativity with the ideas of wave-particle duality and the Uncertainty Principle at the heart of quantum mechanics. This would, in effect, be the "theory of everything" which the string theorists have been looking for.

Stephen argues that, even without having found a convincing, precise theory, we can predict what some of its effects would be. So he can begin to

---

Stephen Hawking has been honored in unconventional ways as often as in the more traditional ones.

Hawking, leading physicist and author, is honoured

STEPHEN HAWKING, who crowned numerous scientific achievements with the seemingly impossible one of turning theoretical physics into a bestseller, made a Companion of ...... the Queen's ......

By David Li......

and gardener at Buckingham Palace and the postman at Windsor Castle receive Royal Victorian Medals.

The chairman of Independent Television News, David Nicholas, is given a knighthood for his work in adapting the news organisation to the new-technology revolution.

The musical director of the English National Opera, Mark .....er, receives the CBE, as doe.. former Football Association ..etary Ted Croker, who retired ..r this year.

.. new life peers are Profes.. .n McColl, the director of ..gical unit at Guy's Hospi.. ..don, Sir Eric Sharp CBE, ..rman and chief executive .. and Wireless, and Si.. ..ton, an in..

Hawking in Star Trek

STEPHEN HAWKING has been given a cameo role in Star Trek: The Next Generation after confessing he is addicted to the TV show.

The brilliant astrophysicist, who is crippled with motor neurone disease, was visiting the Paramount set in Hollywood where the series is shot ....hen he told studio bosses his .... in an

Producers set about fulfilling his wish with a scene where the android crew member Data programmes the starship Enterprise's computer to create a poker game between himself and Professor Hawking, Einstein and Isaac Newton.

Professor Hawking, author of the best-selling A Brief History of Time, is confined to a wheelchair. He is unable to speak but communicates .... a voice synthesiser.

put together a kind of quantum cosmology: a set of quantum conditions from which the universe could grow.

But if we are to find a dynamic formula which allows this whole process to get started, it is impossible to rely on the conventional mathematics of time and space; we already know that the laws governing them break down at the point of singularity from which they predict the universe must begin. So another way of working the equations has to be developed.

Stephen used a principle called "sum over histories" to work out the best approach to take. It was an approach suggested by an American physicist, Richard Feynman. Basically, it means looking at all the possible ways for something to behave; and then by reviewing them, eliminating the least likely. You then have what should be the most probable way to find the right answer. It is a bit like trying to work out how a letter posted in London might have reached its destination in New York. It could have gone on a direct flight from London to New York; it might have gone on a flight from London to Boston, or London to Washington before being put on a connecting flight to New York. It is even possible that it went on flights to Moscow or Tokyo before ending up in New York. By considering all the possible routes, making a "sum over histories" of all possible routes for a letter going from London to New York, you can begin to see which are the more likely possibilities. The Moscow and Tokyo routes for the letter, for example, should be eliminated as being among the least probable pretty early on. So you end up giving serious consideration to probable routes only.

In terms of finding the best mathematical "tool kit" for defining the early universe, the "sum over histories" approach ends up suggesting something which may sound dubious at first, if we are thinking too conventionally or conservatively. It emerges that the most probable way to work the equations is to consider time in a way developed by Richard Feynman to make it mathematically more flexible. It is called "imaginary" time; but who is able to say with conviction what form time really does take? There is no scientific reason to suppose that "imaginary" time cannot possibly ever exist.

It is a bit like being able to use negative numbers in maths. In the "real" world, you cannot have fewer than no eggs in a box; but in a mathematical world, you can understand the result of having minus two eggs in the box. Add four eggs, and you end up with two in the box. By using "imaginary time," Stephen was able to to organize almost all the essential ingredients for an explanation of the early universe. And he was able to consider and compare the result with the likeliest "real" time ideas.

Looking at all the most probable models, essentially Stephen was able to show that there were three different possible versions of what happened. The first was to have a "real" time universe which starts with a Big Bang. This has the problem we have already acknowledged; it must involve the awkward singularity at the beginning. The second option, also a "real" time model, suggests that the universe has existed for ever. But this is not without problems either. For a start, you have to explain how Einstein's version of space and time come to exist in such a model. But the third option, an "imaginary" time universe which has existed forever, gave Stephen a much more fascinating picture. It had one or two very useful consequences. As long as Stephen described the universe within the framework of imaginary time, no singularities arose. Everything essential to the development of the universe, including space and time as we understand them, could be curved right round and contained within the primordial universe. There would be no beginning or end to the universe. There were no initial ingredients which needed to be created, like Lemaître's primeval atom. And, by avoiding this, Stephen Hawking avoided the problem of the singularity and the breakdown of the laws of physics.

There was one other elegant bonus. By setting a particular restriction on the nature of the universe, Stephen was able to demonstrate that only a universe very like our observed universe could exist according to this "imaginary" time theory.

A boundless universe

Stephen's model is known as a "no boundary" model of the universe. He realized that the universe would need to have no boundaries to ensure that only our universe would evolve from the uncertain quantum nature of the early universe. Left unrestricted, all universes are equally possible from the kind of quantum start Stephen proposed. Intellectually, this would be no real improvement on Andrei Linde's second version of inflation, which produced an infinite number of possible universes. Our universe had to be one of the possibilities; but the theory failed to predict how or why it would become the one we are in. Stephen argued that this problem could be avoided if the quantum universe evolved into a "no boundary" universe. This essentially means that there are no boundaries marking the end of space or time, even though the universe as a whole has a finite size.

It may be a little easier to grasp this rather difficult idea if you imagine walking across the surface of a sphere, like the Earth. However far you walk, and in whatever direction you go, you never meet a boundary to mark the end of anything. You can simply keep going forever, around and around the surface. And, if you think about it, the same would apply if—instead of the Earth—you were walking around the surface of a giant balloon. In fact you could be walking either on the outer surface of the balloon, or inside the balloon on its inside surface. In other words, the "no boundary" universe does not have to be any specific shape or size; it simply has to give space and time the kind of continuity without boundaries that the inside or outside of a balloon offers.

There are mathematical reasons why, if the universe has this "no boundary" characteristic, it is the most probable outcome from the kind of quantum start which Stephen Hawking has suggested to explain the first moments of the life of the universe. And the "no boundary" proposal, also very impressively, can embrace a number of other theories, like inflation theory, which seem very likely to explain some aspect of the early universe.

So it has a lot of attractive features. But it has not, of course, been any more proven than the other purely theoretical ideas which have been proposed as explanations of the early universe. Stephen Hawking himself insisted in *A Brief History of Time* that it is only a proposal which cannot be deduced from some other principle. In the end it is just a workable possibility developed by Stephen in collaboration with Jim Hartle at the University of California.

However ingeniously he applied various principles and laws to his cleverly constructed idea, Stephen also knew that there was one problem he could not avoid. He had to find a way for gravity to be subject to quantum laws; and no one has yet come up with a satisfactory quantum theory of gravity. Even if he could manage to avoid most of the difficulties he had posed with his own singularity theorems, he could not do so without having a theory of everything to unite gravity and quantum mechanics.

However, assuming that there will eventually be a satisfactory quantum theory of gravity, one of its consequences in Stephen Hawking's model of the universe is that there will be tiny fluctuations in the earliest moments of the universe, because the Uncertainty Principle will allow them to occur in a vacuum. And then, as inflation theory allows the primordial universe to expand, these tiny fluctuations will become the differences in temperature which COBE detected in the background radiation from the Big Bang. So, for Stephen Hawking, COBE's results were a confirmation that his model of the universe was at least feasible.

Critics of the proposal suggest that all Stephen has done is to find an ingenious way of circumnavigating the essential difficulties of reconciling relativity and quantum mechanics. Or they point to the fact that pivotal to its effectiveness is a quantum theory of gravity which no one has yet discovered; and for that we may have to wait until something emerges from string theory. And, even if that is sooner rather than later, few physicists are suggesting that this will be the last chapter written in the book of physics. Ed Witten made the perceptive comment that any theory rich enough to be able to give us a theory of everything is bound to be rich enough to throw up a whole set of fresh challenges when it does so.

Where we are now

Thinly veiled by Ed Witten's comment is an apparent admission that we may never be any nearer, or further from, a complete understanding of the universe than we are now. The suggestion is that we are close to knowing everything, but we may never actually get to the point where we know everything. The more we peel back the layers of the onion, the more layers we may find. The final solution to the mystery may be as elusive as it is hard to define. But, even if we still seem baffled about the fraction of a second in which the Big Bang came to life, most cosmologists are extremely bullish about the progress they are making—including Stephen Hawking. It is, after all, only a fraction of a second we still have to explain. And we have come an extraordinarily long way from the early idea of a lid over the Earth with holes in it, through which firelight shone as the stars. Our understanding of the universe is a major triumph for careful scientific observation and inspired theoretical reasoning. And Stephen Hawking's proposed model of the universe differs in only the tiniest details from the one practically every other cosmologist would now piece together.

String theory may soon give us an equation which explains how energy popped up in a vacuum and then, according to inflation theory, expanded very rapidly. As it did so, some kind of quantum fluctuations in the vacuum would have ensured that there were tiny irregularities in the Big Bang explosion it triggered. These irregularities would appear as differences in temperature of no more than 0.002 of a degree some 300,000 years later when they were detected by the COBE satellite. But they were still sufficient to allow matter to evolve in an uneven fashion as the huge heat from the Big Bang started to cool.

This is how it could have happened. From the split second after the Big Bang, the energy from the explosion started to spawn primordial particles of matter such as quarks. But they did not simply develop as one step in an irreversible process. In the first second of the life of the universe they would have collided and returned to pure energy and then turned back into particles once more. They would have done this several times as part of a crazy dance of creation in an impenetrably hot plasma, eventually producing the kind of interactions from collisions that we have observed in the most powerful particle accelerators. Within three minutes, more substantial structures would have begun to take shape, as the first nuclei which were destined to make atoms started to form. The heat was beginning to cool just enough for the earlier particles to begin to bind together. But it was still outrageously hot and it was impossible to see into the white hot cauldron of the emerging universe.

We have to wait 300,000 years before the universe finally clears, and electrons start to encircle the nuclei to form atoms—roughly 80 percent hydrogen and 20 percent helium. It is a billion years after the Big Bang before gravitational pressures have created the first stars, quasars and the jostling clumps of matter which are turning into galaxies. In the first stars, hydrogen fuses to make

Stephen Hawking

helium and the stars begin to shine. Black holes in the center of quasars exert huge gravitational pulls on matter all around them; the swirling discs of matter they create get hotter and give birth to new stars. Older stars are drawn towards the vast clumps of dark matter gathering around the bright quasars. They perhaps fall into these giant rotating wheels, settling into the spiral arm patterns we are now familiar with.

As the earliest stars come to the end of their life cycles, the first white dwarfs are formed and gradually cool to become invisible brown dwarf stars. The earliest large stars implode and supernovas appear, creating neutron stars and spreading all the heavier elements across the universe. Our star, the sun, is born, and its planets are drawn around it by gravity. Eventually the Earth is cool enough for life to evolve; and human beings finally emerge—just one of the extraordinary products to develop from the dust of stars. It is not long before they begin to discover clues which lead them to understand almost every detail of this unbelievable chain of events, some 15 billion years after it all began.

Knowing the mind of God

It might seem all too easy to make the same mistakes as earlier generations, and assume that our present view of the universe has to be the ultimate explanation. After all, in their day, the Ptolemaic and Newtonian models of the universe also looked unquestionably right. Compared with them, the Big Bang model of the universe is still in its infancy. But these earlier mod-

---

The Horsehead Nebula in the constellation of Orion is one of the most photographed beauty spots in the universe. The Horsehead is a distinctive dark dust cloud at the center right of the picture.

els of the universe were proved wrong by dramatic steps forward in what we could learn from observation. Ptolemy and the cosmologists of his time used incredible ingenuity to construct a working model of the universe from observations with the naked eye alone. Galileo's telescope showed us more details of the solar system and proved Ptolemy's model wrong. Newton developed a theory of gravity which made sense of the observed solar system 300 years ago; then Hubble's observations beyond the solar system revealed the static shortcomings of Newton's infinite and eternal model. And, in the comparatively short time since Hubble's observations, we have put telescopes in space and observed as far back in the history of the universe as we know it is possible to look. It is not as if there is much scope for improved observation to show us anything more; it seems that we have now seen almost all of the visible universe.

With so much knowledge of our universe, it may not seem too immodest to say that we are close to understanding what it is and how it works. Even though perhaps 90 percent of the universe remains undetected, since we have yet to discover all the dark matter, we can deduce enough about it to understand its role. Until we do detect it and know what its total mass is, the ultimate fate of the universe must remain a mystery. But, even here, we have a clear idea of the possible alternative endings.

But this is far from saying that we understand almost everything about the universe. Knowing its dynamic history does not even begin to address some of the eternal philosophical questions, such as why is the universe here? Or, what is it all for? The principles of science alone may not be able to answer these questions. But Stephen Hawking is confident that progress in cosmology can only improve the chances of us finding good answers to such questions. Knowing how the universe works must help inform our thinking on why it exists, whether or not it was created, and whether or not it has a purpose. At the end of *A Brief History of Time* Stephen concludes that, if we do discover a complete theory of everything, its basic principles and implications should in time be understandable by everyone.

And, once we all understand the true nature of the universe, we can all take part in the discussion about why it exists. Should we ever resolve that question, he suggests, it will be "the ultimate triumph of human reason—for then we will know the mind of God."

Perhaps, for many of us, that challenge will seem a step too far. There are millions of us who have never before got close to discovering the nature of the universe. We may just not have tried; more likely we were convinced that it was beyond our limited capacity to understand. Simply to know the nature of the universe, as Stephen Hawking and cosmologists throughout history have been able to understand it, may give us as complete and as satisfying a context for our lives as we need.

Illustration Credits

Picture research by Frances Topp and Vivien Adelman.
Artwork on pages 20, 22–23, 154–55 and 240 by Colin Pilgrim.

Ace 245; Advertising Archives 129 (right); AIP Emilio Segrè Visual archives 63, 82, 97, 205; Ancient Art and Architecture Collection 117, 119; Astronomical Society of the Pacific 161; BBC Picture Archives 87; BFI 219 (top); Birr Castle 58; Bridgeman Art Library 26; Giraudon 27, 28, 35, 38, 44, 45; British Library 44; Caltech 150 (top), 197, 227; Camera Press 177, 178, 254; Carnegie Institution of Washington 72, 73; Cavendish Laboratory 137, 138, 139; Corbis 26, 129, 150, 176, 203; Cordon Art 215; ET Archive 36, 51; Mary Evans 39 (top); David Filkin 98, 111, 123, 181, 187, 252, 261, 266; John Frost 172, 271; Genesis Picture Library 91; Huilton Getty 20, 34, 85, 99 (left), 206; Illustrated London News 128; Image Bank 25, 65; Kobal 219 (bottom); Dmitri Linde 240; Magnum 202; Manni Mason's Pictures 5, 8, 101; NASA 110; National Trust 49; Network Images 243; NHPA 253; Novosti 225; Roger Penrose 102; PhotoScala, Florence 39 (bottom), 41; Pictor 235; Popperfoto 126, 177; Justin Pumfrey/©BBC 277; St. John's College, Cambridge 148; Science Museum 47, 134, 137, 164; Science Photo Library xii, 2, 12, 19, 30, 33, 37, 42–43, 48, 50, 54, 56, 57, 58, 59, 65 (inset), 68, 71, 76, 79, 81, 88, 94, 99 (right), 103, 109, 110, 112, 114, 121, 124, 125, 127, 132, 134, 140–41, 147, 158, 162, 166–67, 168, 169, 174, 183, 184, 189, 192, 194, 195, 200–1, 209, 210, 212, 214, 218, 220, 223, 226, 232, 238, 244, 250, 262–63, 268, 279; Tony Stone 15; Telegraph Colour Library 122; University of California 108, 190.

Index

Index

energy inflating into vacuum, *see* inflation theory

epicycles, 26, *27*, 34

Eratosthenes, 18–21, 70; demonstrates curvature of Earth's surface, 18–20; measures circumference of Earth, 21

Escher, M.C. (artist), 215, *215*

exotic dark matter, 175–90; neutrinos, 175–86; WIMPs, 14, 187–90

expanding universe, 171; detected by Hubble, 72–74, 84; expansion counterbalances gravitational attraction, 101; indicated by Einstein's mathematics, 83; projected back in time to primeval atom, 86, 111–12; quasar recession speeds, 196–98

extraterrestrial intelligence, 194–96, 230–31

Faber, Sandra, *190;* three-dimensional movement plots, 188–90

Feynman, Richard, 272

Filkin, David (author), 4–6, *5,* 9–11

Fraunhofer, Joseph von, 58–62; identifying elements in laboratory, 130

Fraunhofer lines, 58–62, *61, 65,* 73, 98, 171; caused by absorption or emission of light, 61; Doppler shift observed in star light, 67–69, *68,* 74; Huggins discovers significance of, 61–62; identification of sun as a star, 61–62; in light from quasars, 197–98; stellar compositions, 61–62, 74; supernovae, 92

Frenk, Carlos, 184–86, *185*

fusion reactions within stars, 89–92, 156, 277–78

galaxies, 56–57, *56–57, 94, 188, 200–201, 268;* black holes at center of, *218, 223,* 227–28, 229; countercurrents of movement, 190; Doppler shift effect, *63;* Great Andromeda Galaxy, *158;* haloes of dark matter, 164, 169–70; Hubble classification, *74;* individual stars observed within, 57; jets of matter, *114, 200–201, 212, 223, 226, 226,* 229; Large Magellanic Cloud, *xii,* 169–70, *166–67;* mapped by Messier, 57; red shifts, *65, 68,* 74, 75, 112, 171; rotation pattern explained by dark matter, 278; speeds of recession increase with distance, 72–74; slow with time, 171; Whirlpool Galaxy, *59, 71*

Galileo (Galileo Galilei), 38–47, *39;* confirms sun-centered orbit of Venus, 44; conflict with Church, 44–47, *44, 45;* Inquisition imposes house arrest, 46; Jupiter's moons, 41–44, *42–43;* moon observations, *41;* popularizes sky watching, *41;* renounces Copernicanism under duress, 45; study of acceleration under gravity, 40, 52; telescopes, *39, 41,* 280

gamma rays, 139

Gamow, George, 97–99, *97*

gravitational lensing, 165–70, *166–67*

gravitational pressures; drive atomic fusion in stars, 89–90, 156, 277; lead to inevitable collapse of matter, 214–15; trigger supernovae in larger stars, 92

gravity; Albert Einstein's model, 80–82; effects studied by Galileo, 40; impor-

tance in evolution of universe, 271, 275; Isaac Newton's model, 47–52, *48,* 80–82, 252; quantum theory of, 224, 271–72, 276; rubber sheet analogy, *81,* 81–82; slowing the expansion of the universe, 171–73

Guth, Alan, 243–44, *244*

Hawking, Stephen, 101, *101, 271, 277;* admiration for Galileo, 10, 38–39, 47, 104; Andrei Linde acts as interpreter, 246–47; bet with Kip Thorne, 204, 227–28; black hole emission of particles, 228–29; cox of University College Rugby Eight, 4–6, *5;* and David Filkin, 4–6, *5,* 9–11; Lucasian Professor of Mathematics, Cambridge, 7; model of Big Bang, 104, 216–17; Motor Neuron Disease, 6; scenarios for development of universe, 271–76; voice synthesizer, 6, *8, 9; see also A Brief History of Time*

Heisenberg, Werner, *see* Uncertainty Principle

Herschel, William and Caroline, discover Uranus, 57

Hipparchus discovers planets (wandering stars), 24

Hooker Telescope, *72–73*

horn antennas, *98, 99,* 100, *111*

Horsehead Nebula, *279*

Hoyle, Fred, 87–89, *87,* 96; life cycle of stars, 87–92, 96; steady state theory, 87–92, *87*

Hubble, Edwin, 69–74, *72, 85;* classification of galaxies, *74;* expansion of the universe, 72–75, *76,* 85; galactic distances from Earth, *71,* 72–74; red shifts in galactic spectra, 73–74, 112; relation

Hubble, Edwin (*cont.*)
between galactic speeds and distances, 75; *see also* expanding universe; galaxies; red shifts
Hubble Space Telescope images, *91, 168, 189, 218, 232*
Hubble's Law, 75
Huggins, William, 61–62, 64

imaginary time, 272–73
inflation theory, *235,* 241, 242–44, *244, 245,* 276; multiuniverse version, 246–47, 265–66

jets of matter, *200–201, 210, 226;* from binary systems with a black hole, 226–27; from galaxies, *114, 196, 223,* 226–27, 229
Jupiter, moons of, *42–43*

Kant, Immanuel, 56
Kepler, Johannes, 33–38, *35;* assistant to Tycho Brahe, 37; concept of a force acting at a distance, 34; elliptical orbits, 34, *35–36,* 37, 51–52; journey to visit Tycho Brahe, 36; magnetic attraction hypothesis, 34

Large Magellanic Cloud, *xii,* 169–70, *166–67*
Leavitt, Henrietta, 70
Leeuwenhoek, Anton van, 40
Lemaître, Georges, 77–78, *82,* 83, 104, 199; and Einstein, 84–85, 208; moment of creation, 84–85, 86; primeval atom *see* primeval atom; rejects Einstein's cosmological constant, 84; supported by Catholic Church, 85
light deflected by gravity, *168;* predicted by Einstein,

164–65; proved by Eddington's experiment, 165
light years, 67
Linde, Andrei, *240,* 241–49, *243,* 251–52, 274
Lucasian Professor of Mathematics, Cambridge, 17, 47, 148

MACHOs (Massive Astrophysical Compact Halo Objects), 14, 164–65, *166–67,* 173, 187, 190; revealed by gravitational lensing, 165–70, *168;* search in Large Magellanic Cloud, *166–67,* 169–70
magnetic attraction between bodies (Kepler), 33–34
Mars orbit; Kepler shows to be ellipse, 37–38; perfect fit to Newtonian theory of gravity, 51
matter; excess over antimatter, 153; non-existent in extreme conditions of initial Big Bang, 236; orbiting black holes, 217–19
matter collapsing under gravitational forces, *103–4,* 199, 202, 214–17; drives atomic fusion in stars, 89–90, 156, 277–278; inevitable consequence of Einstein's mathematics, 101–4, 216; leading to black holes, 205; reversed in time to model Big Bang, 106–7, 214–16
Mendeleev, Dmitri, 121, *121,* 123–25, 130, 131
Mercury orbit; agreement with Einstein's theory of gravity, 82; discrepancy with Newton's theory of gravity, 48–52
Messier, Charles, 57
models of the universe; Babylonian, 14, 17; Big Bang *see* Big Bang; Copernican,

32–33, *33, 35;* dynamic v. static, 83; Newtonian, 52–53, *50,* 83, 111, 278–80; Ptolemaic, 26–29, *27, 32, 33,* 40, 111, 278–80; Pythagorian, 21–23; steady state, 87, 95; theoretical computer generated, 265–67
moment of creation, 92–93; concept rejected by Einstein, 85–86; disliked by atheistic scientists, 92–93; proposed by Lemaître, 84–85; *see also* Big Bang; primeval atom
moon, observed by Galileo, *41*
Mount Wilson Observatory, 70, *72, 73,* 85, *85*

NASA, 109
nebulae, *2, 50, 56, 76, 174, 279*
neutrinos, *163,* 176–86, *181,* 190; detector built by Frederick Reines, 177–80; possible candidate for dark matter, 181, 182, 184, 187; predicted by Wolfgang Pauli, *176,* 177; tested for decay, 180–82
neutron stars, 92, 277
Newton, Sir Isaac, 7, *47;* gravitational theory, 47–51, *48,* 80–82, 280; laws of motion, 48, 52–53, 80; Lucasian Professor of Mathematics, Cambridge, 47; planetary orbits explained by gravitational attraction, 48–51; *Principia Mathematica, 49;* reflecting telescopes, *51, 55;* universe infinite in space and time, 52–53, *50,* 83–84; *see also* gravity; models of the universe

Oppenheimer, Robert, 199, 202–4, *202, 205*
Orion Nebula, *ii, 50*

Parsons, William; confirms individual stars within galaxies, 57; first observer of spiral galaxies, 59; Leviathan telescope, 57, 58

particle accelerators, 142–48, 151–53, 238; CERN, 140–41, 147, 237; create temperatures close to initial Big Bang, 152, 236, 238; exotic particles, 173; limits on size, 237–38

particles created from pure energy, 151–53, 154

Pauli, Wolfgang, 176, 177

Penrose, Roger, 102–4, 102, 103, 204, 214–16, 214

Penzias, Arno, 99, 100, 105, 108

Periodic Table, 123–24, 123, 131

photons, 221–22

pitchblende, native ore of uranium, 128

Planck Explorer project, 264–67, 266

planetary orbits; circular, Earth-centered (Copernicus), 32, 33; circular, Earth-centered (Ptolemy), 26–27, 26; circular orbits favored by Greeks, 20–24, 25, 26–29; elliptical, sun-centered (Kepler), 32–33, 51–52; epicycles, 26, 27, 32–33, 34; reverse movements in sky, 26

planets; change in brightness, 21–24; formation of, 156

Plato, 24

primeval atom, 84, 93, 111, 117, 208, 273; corresponds to singularity, 104; possible source of hydrogen in Big Bang, 96; postulated by Lemaître, 84

Ptolemy, 26–27, 27

pulsars, 92

Pythagoras, 21–23, 116; harmony of the spheres, 21

quantum mechanics, 187, 222–24, 255, 271–72; applied to vacuum laws, 242; mismatch with relativity, 255–56

quantum theory of gravity, 224, 270, 274–76

quarks, 256, 277; see also string theory

quasars, 168, 196–203, 204, 208, 219, 231, 236; early features of galaxy formation, 229, 277; in hearts of galaxies, 229–30; speed of recession, 198

radio signals, 193–96, 200–201; from galaxies in collision, 196

radioactive decay; conversion of mass to energy, 144–45; products detected by Rutherford, 139, 142

radioactivity, 105, 128, 129; commercial applications, 128, 128, 129; discovered by Henri Becquerel, 125–26, 125; forms of radiation, 138, 139, 139, 142; harmful effects, 130; studied by Marie and Pierre Curie, 126, 127–30, 127

radium, 127, 128, 129; glows in the dark, 129, 130; isolated by Curies, 130

red shifts in spectra, 61; discovered by Hubble, 72–74; from binary systems, 67–69, 68; from galaxies, 65, 68, 72–75, 112, 171; in light from quasars, 196–98

Reines, Frederick, 177–80, 178

Relativity, Einstein's Theories of, 78–83

rubber sheet analogy of space-time, 80–82, 81

Rubin, Vera, 159–61, 161, 170, 184

Rutherford, Ernest, 133–36, 135, 139, 142–43; forms of radiation, 138, 139, 139, 142; gas chamber experiment, 134–36; particle accelerator, 142–44, 151–52; radioactive decay products, 134–36, 139, 142

satellites, 250, 262–63; COBE, 105, 109, 109–11, 110, 265; Planck Explorer, 264–67

scalar field theory, 248

Schmidt, Maarten, 197–98, 197

Sciama, Dennis, 96, 101, 102, 204

SETI (Search for Extra-Terrestrial Intelligence) Institute, 196

singularities, 102, 103, 220–24, 231; breakdown of laws of physics, 102, 104, 215–16, 270, 271, 273; do not arise in framework of imaginary time, 273; inevitable result of black holes, 102, 103; origin of Big Bang, 104, 111–12, 221; possible end of universe, 171; result of gravitational collapse of matter, 214–16; in scalar inflation theory, 248; survival of the fittest, 252–53

Smolin, Lee, 252–53, 252

Smoot, George, 105, 108–11, 108, 111; COBE satellite see COBE; data-gathering techniques, 108–9, 108; variations in background radiation, 105, 109–11, 112, 270

spectrum of refracted light, 58–62, 66; see also Fraunhofer lines; red shifts in spectra

Starry Messenger (Galileo), 44

stars; binaries see binary star systems; Cepheid variables,

Index